Lecture Notes on Coastal and Estuarine Studies

Lecture Notes on Coastal and Estuarine Studies

Managing Editors:
Malcolm J. Bowman   Richard T. Barber
Christopher N.K. Mooers   John A. Raven

18

Flemming Bo Pedersen

Environmental Hydraulics: Stratified Flows

Springer-Verlag

Berlin Heidelberg New York London Paris Tokyo

Author

Flemming Bo Pedersen
Institute of Hydrodynamics
and Hydraulic Engineering
Technical University of Denmark
Building 115
DK-2800 Lyngby

ISBN 978-3-540-16792-1   ISBN 978-3-642-86600-5 (eBook)
DOI 10.1007/978-3-642-86600-5

ʼbrary of Congress Cataloging-in-Publication Data. Pedersen, Flemming Bo, 1939- Environ-
ʼtal hydraulics. (Lecture notes in coastal and estuarine studies; 18) Bibliography: p. Includes
    1. Stratified flow. 2. Hydraulics. I. Title. II. Series: Lecture notes on coastal and estuarine
    18.
      ʼ7 1986 627'.042 86-20247

            Heidelberg 1986
            ʼdcover 1st edition 1986
                ʼs Beltz, Hemsbach/Bergstr.

CONTENTS

(cont.)

PART III

CASE STUDIES

(cont.)

V

─────

Page

SUMMARY

The present lecture notes cover a first course in the most common types of stratified flows encountered in Environmental Hydraulics. Most of the flows are buoyancy flows, i.e. currents in which gravity acts on small density differences.

Part I presents the basic concepts of stagnant, density - stratified water, and of flowing non-miscible stratified fluids. The similarity to the (presumed) well-known open channel flow, subject to a reduced gravity, is illustrated.

Part II treats the miscible density stratified flows. In outlining the governing equations, the strong coupling between the turbulence (the mixing) and the mean flow is emphasized. The presentation and discussions of the basic governing equations are followed by illustrative examples. Separate chapters are devoted to Dense Bottom Currents, Free Penetrative Convection, Wind-driven Stratified Flow, Horizontal Buoyancy Flow and Vertical jet/plumes.

Part III presents some examples of practical problems solved on the basis of knowledge given in the present lecture notes.

It is the author's experience that the topics treated in chapter 8 and in the subsequent chapters are especially well-suited for self-tuition, followed by a study-circle.

ACKNOWLEDGEMENT

The author has benefited by the valuable help of his collegues at the Institute of Hydrodynamics and Hydraulic Engineering, the Technical University of Denmark, especially our librarian Mrs. Kirsten Djørup, our secretary Mrs. Marianne Lewis and our technical draftsman Mrs. Liselotte Norup.

The permission granted by the American Society of Civil Engineers to use two papers from the ASCE Journal of the Hydraulic Division, is highly appreciated.

PART I

NON-MISCIBLE STRATIFIED FLOWS

- CHAPTERS 1 - 3 INCL. -

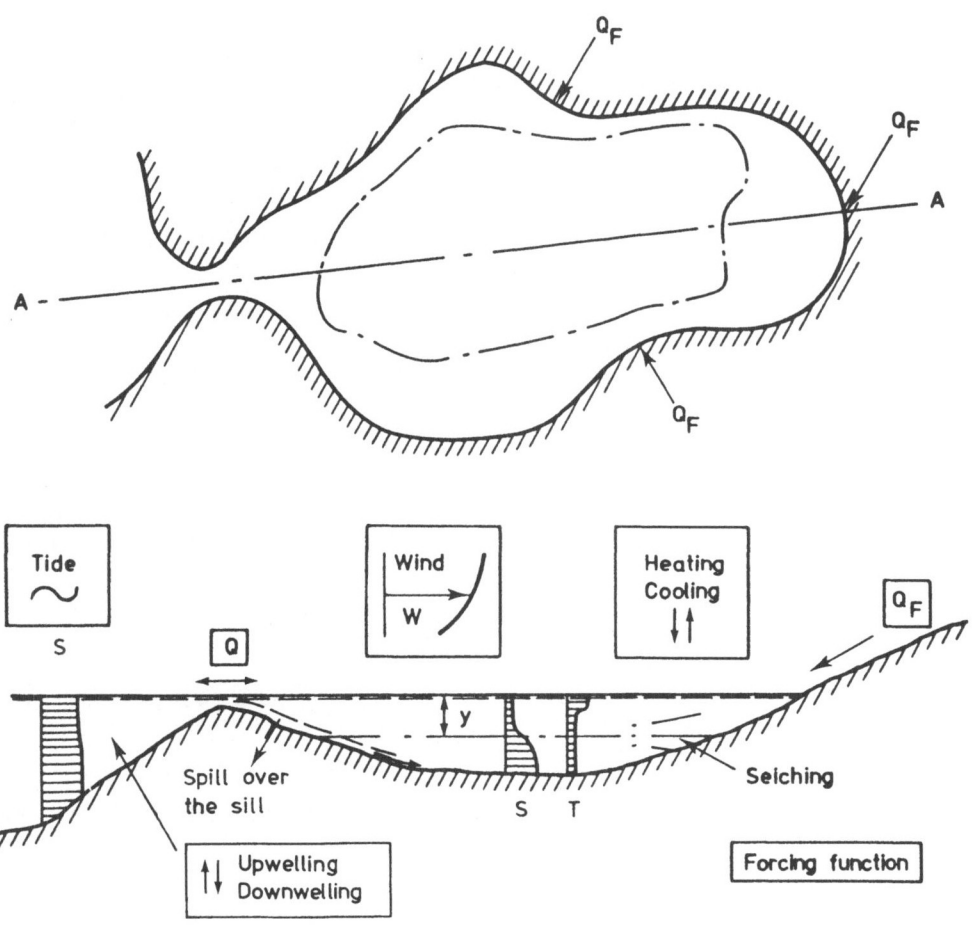

"An estuary is a semi-enclosed coastal body of water having a free connection with the open sea and within which the sea water is measurable diluted with fresh water deriving from land drainage". Pritchard [1967].

1. INTRODUCTION

The explosive economic growth, which was initiated in the early 60's, has left its mark on the hydraulic engineering design, evaluation, and advise in a wide spectrum of environmental problems. Especially, the Scandinavian countries with their many island, fjords, and lakes offer a great challenge to the hydraulic engineering of lakes, reservoirs, coastal and estuary hydraulics with respect to

Pollution –   (domestic and industrial waste water, cooling water, oil spill, agricultural manuring etc.)

Water management – (drinking water, industrial water, irrigation, hydropower etc.)

Navigation –   (currents, ice formation, shoaling, erosion-involving dredging, protection works etc.)

Floods –   (forecasting, dikes, regulation of discharges)

Traffic –   (tunnels, bridges, harbours etc.)

Offshore activities – (coastal currents and eddies, ice drift, blow-outs etc.)

Fishing –   (currents, salinity, temperature, oxygen content, formation of fronts).

None of these problems have a single solution, because many aspects – which do not follow the conservation laws of nature – will affect the process of decision in different directions. However, it is necessary to make the final (often political) decision on as good a technical basis as possible. This requires the civil engineer to be able to predict the effects of the interference with the lake, the estuary etc., which means that she/he must be able not only to prove that the primary goal is fulfilled, but also to point out the possible secondary effects.

The civil engineer's part of the job will be to determine the time and space variables of

water level

density

velocity

transport of matter, and

water quality.

Most of the problems faced in the natural aquadic environments are complicated by space and time variations in the density stratification, either caused by temperature or by salinity effects (or suspended particles). The formation of thermoclines (vertical temperature jumps), haloclines (vertical salinity jumps), and pycnoclines (vertical density jumps) has a great influence on the water circulation and the water quality. For instance, a halocline is formed where the fresh water from a river enters the sea. The fresh water - which is lighter than the ocean water - forms a light upper layer, which acts as a lid on the ocean water. Suppose that the two water bodies couldn't mix. Then a stagnant lower layer of ocean water, suffering from a lack of access to the oxygen in the air above, would soon become an aquadic desert with no higher life of flora and fauna. Fortunately, the fresh water and the sea-water are miscible, which means that a steady or intermittent transport of lifegiving oxygen takes place - in one way or another - from the air to the ocean, even to the deepest parts.

Mixing of two fluids which form a stable stratification (for instance fresh water above salt water) demands an energy supply. In nature, this energy originates from many sources, such as wind, tides, heat exchanges, evaporation, ice formation, barometric pressure variation, gravity etc. The manor part of the energy supply is used for other purposes than mixing. For instance, the energy transferred from the wind to the water is mainly used to create waves, circulation, setup (upwelling, downwelling in the ocean) and hence to produce turbulent kinetic energy, which eventually is dissipated into heat. When the wind ceases, the waves, the flows, and the setup die out, but due to

the mixing, part of the dense salt water has been lifted against
gravity and mixed with the fresh upper layer, and hence it has
gained potential energy. The efficiency of this mixing process -
i.e. the ratio between the gain in potential energy and the energy input (the production of turbulent kinetic energy) - called
the flux Richardson number - has appeared to be a constant in a
great number of density stratified flows. Hence, we shall devote
some effort to estimating the gain in potential energy due to
mixing and to the production of turbulent kinetic energy, both
terms appearing in the energy equation for the turbulence. Later,
examples are given, in which the flux Richardson number is obviously not a constant. These examples call for a more sophisticated theory, which includes the gain in turbulent kinetic energy of the entrained water.

The mathematical description of the physical processes in
inhomogeneous fluids are based on the following set of equations

The equation of mass conservation

The equation of continuity

The equation of motion

The energy equation for the mean motion

The energy equation for the turbulence.

In order to solve a specific problem, some idealization
must be introduced. Before this step is taken, we must realize
that the more detailed we want to describe the phenomena, the
more difficult it will be to formulate the boundary conditions
and to specify the necessary empirical constants and/or functions. Therefore, the model should not be too complicated, but,
on the other hand, it should give sufficient information to be
applicable to the engineering problem. One large class of buoyancy flows in geophysics may be idealized to two-layer flows
with an active, turbulent flow and a passive, non-turbulent ambient fluid, and hence we shall pay special attention to this
kind of flows.

In part I - which primarily deals with non-miscible density stratified flows - we gain some physical insight into the ma-

jor buoyancy effects on the mean motion, namely that the flow behaves like an ordinary (homogeneous) flow exposed to a highly reduced acceleration of gravity.

Part II is primarily dealing with miscible buoyancy flows. The mixing affects the governing equations for the flow, i.e. the flow field and the stratification depend on the mixing. In turn the mixing depends on the flow and the stratification. This strong coupling between the turbulence (the mixing) and the mean flow is discussed from a physical point of view, as it is of major importance for a basic understanding of miscible buoyancy flows. In the presentation and discussion of the governing equations, a number of simple examples are used. The last of part II of the lecture notes is devoted to a number of geophysical phenomena as well as to some man-made buoyancy flows. The examples have been presented in "self-contained" chapters, and hence - some of the chapters may be deleted in a course, without diminishing the profit of the other chapters. Similarly, it is not necessary to follow the present order of the examples.

Finally, in part III some case studies are presented. One example discusses the influence on the Baltic Sea and the Danish inland waters (the Cattegat, the Belt) for a reduction of the river runoff to the Baltic. The example illustrates clearly that a local impact has a global effect, - a very common phenomenon in estuary and coastal water managements. Most of the examples given have been worked out in cooperation with the author's master and doctor students.

2. PRESSURE CONDITIONS AND POTENTIAL ENERGY

From the basic course in hydraulics we know that the pressure conditions in a stagnant pool of water is hydrostatic, which means that the pressure equals the weight of the fluid column above it. On the assumption that we are concerned with the pressure normal to the flow direction, the pressure distribution is hydrostatic here too, but now the pressure balances the vertical component of the weight of the fluid column normal to the flow direction. Contrary to homogeneous fluids, this weight generally is a function of space and time due to the mixing of water masses of different temperature, salinities, concentration of suspended particles etc. The density of sea water is discussed in Appendix (read it!).

The pressure term can be found in all the equations of motion, not as the absolute value, but as the pressure gradient. Therefore, one can without a loss in generality reduce the pressure by the pressure in a stagnant pool with a density distribution corresponding to the conditions prevailing before the appearance of the density flow, because in this stagnant fluid the horizontal pressure gradients are zero. As far as the vertical pressure gradient is concerned, this is counteracted by the weight of the stagnant water. In the system of reduced pressure, the weight is therefore reduced too - which means that we are dealing with the buoyancy rather than with the weight of the density current fluid. The main advantage of performing this pressure reduction is that the absolute values of the pressure terms then become of the same order of magnitude as the other terms in the equations of motion. Furthermore, as the conditions in the ambient fluid are quasi-stagnant, the two-layered problems are automatically reduced to nearly one-layer problems, in which way it gains in clarity.

Similarly, we often gain in computational accuracy by calculating the potential energy in the mentioned reduced pressure system. The argument for this simplification is that we are concerned with the change in potential energy rather than with the absolute value.

By reducing the pressure and the potential energy with

their respective values in a reference system we make the flow of inhomogeneous immiscible fluids become especially simple to handle, as they turn out to behave like ordinary open channels flow exposed to a reduced acceleration of gravity. This is true for miscible fluid too, but here the reduced acceleration of gravity may be a function of space and time, which will complicate the calculations. Two implications of this reduced acceleration of gravity will be mentioned here and illustrated by examples. First, all gravity driven currents are in "slow motion". Secondly, a surface slope - for instance set-up by wind in a stratified lake - is counterbalanced by a highly exaggerated interfacial slope.

Most mixing processes in inhomogeneous fluids are associated with a change in potential energy. When releasing a heavy fluid on top of a lighter fluid - or similarly, a light fluid at the bottom of a heavy fluid - potential energy is lost and hence available for other processes such as circulation and mixing. Contrarily, mixing of a stable stratification demands an energy supply, as the mixed fluid gains potential energy. As experience from field and laboratory experiments has taught us that these processes are associated with a certain efficiency, we have an effective empirical tool for estimating the mixing, provided we are able to quantify the energy input from the external forcing functions (wind, tide, evaporation etc.) and the change in potential energy as well.

The pressure conditions

We consider a column of fluid in which the density $\rho(z)$ is constant or increasing with depth z, i.e. a stable stratification, confer Fig. 2.1.

As no shear stresses are present in a quiescent fluid, the pressure distribution is determined by

$$\frac{\partial p}{\partial z} = \rho(z) g \qquad (2.1)$$

(where p = pressure, g = acceleration of gravity).

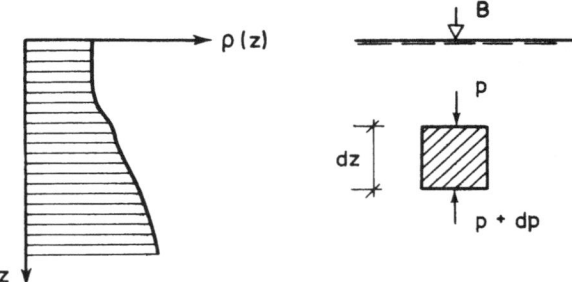

Fig. 2.1 Stable inhomogeneous column of fluid (∂ρ/∂z ≥ 0).

Integrated over a vertical column, equation 2.1 yields the hydrostatic pressure distribution

$$p(z) = B + \int_0^z \rho(z)g\,dz \qquad (2.2)$$

(where B = the barometric pressure).

Contrary to open channel flows, the geophysical extension of stratified flows may be so large within the area of interest, that changes in the barometric pressure have to be taken into account.

Denoting the reference conditions by an index R, we obtain the following pressure distribution in a column of reference fluid at the same location:

$$p_R(z) = B + \int_0^z \rho_R g\,dz \qquad (2.3)$$

The reduced pressure p_Δ, relative to the conditions in the ambient fluid before the introduction of the density flow, then reads

$$p_\Delta = p - p_R = \int_0^z (\rho - \rho_R)\,g\,dz \qquad (2.4)$$

The dimensionless reduced mass $\Delta = (\rho - \rho_R)/\rho_R$ and the reduced acceleration of gravity $g' = \Delta g$ (see Appendix) are now introduced to yield

$$p_\Delta = \int_0^z \rho_R \Delta g\,dz = \int_0^z \rho_R g'\,dz \qquad (2.5)$$

8

which states that the reduced pressure is equal to the pressure in a column of fluid exposed to the reduced acceleration of gravity g'. Note that g' may be a function of space and time in miscible fluids.

<u>Example 2.1</u>

In Fig. 2.2 a two-layer column of fluid is shown, in which we have used the upper layer density as a reference - but we could as well have used the lower layer as reference fluid, as will be done in the next example.

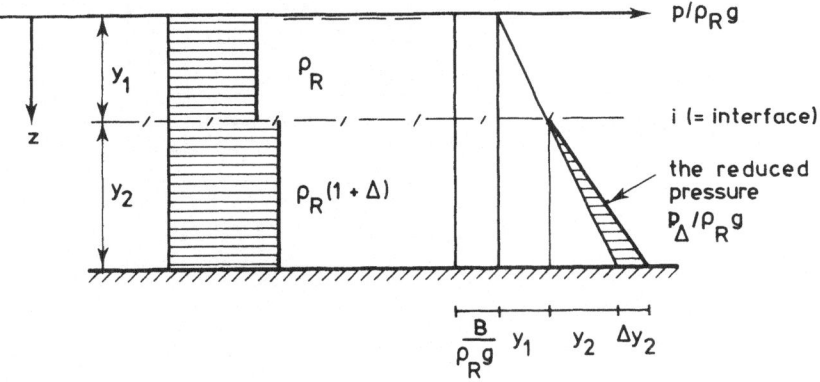

Fig. 2.2   Pressure distribution in a two-layer fluid.

The distribution of the total pressure in the lower layer is

$$\frac{p}{\rho_R g} = \frac{B}{\rho_R g} + y_1 + y_2 (1+\Delta) \frac{(z - y_1)}{y_2} \qquad (2.6)$$

Therefore, the distribution of the excess pressure in the lower layer is

$$\frac{p_\Delta}{\rho_R g} = \Delta y_2 \frac{(z - y_1)}{y_2} \qquad (2.7)$$

i.e. a linearly distributed reduced pressure with the value zero at the interface and the maximum value $p_{\Delta,max} = \Delta \rho_R g y_2$ at the bottom.

The total depth integrated excess pressure is readily calculated to

$$P_\Delta = \frac{1}{2} \Delta \rho_R g y_2^2 = \frac{1}{2} \rho_R g' y_2^2 \qquad (2.8)$$

a result which could have been obtained immediately by use of
the reduced acceleration of gravity concept.

Example 2.2

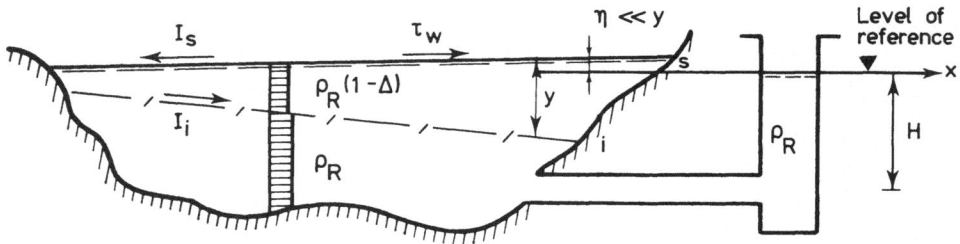

Fig. 2.3   Two-layer stratified lake with wind set-up.

In an initially homogeneous lake a thermocline has been
formed by solar heating, and due to some wind-stirring effects
the upper layer is homogenized, see Fig. 2.3. Experience shows
that almost no shear stress is transferred to the interface, and
hence the bottom layer can be treated as stagnant.

First we look at the superelevation η of the lake surface
relative to the level of reference in the well connected to the
bottom layer of the lake. As the lower layer is quiescent, the
pressure at the level H is the same in the lake and in the well.

$$\left\{ \frac{p}{\rho_R g} \right\}_{lake} = H + \eta - \Delta y \qquad (2.9)$$

$$\left\{ \frac{p}{\rho_R g} \right\}_{well} = H \qquad (2.10)$$

$$\eta = \Delta y \qquad (2.11)$$

The pressure distribution is illustrated in Fig. 2.4. As
the reference pressure is the same all over the lake, it is the

different excess pressures which are to counterbalance the exposed wind shear stress, τ_w, see example 3.3.2.

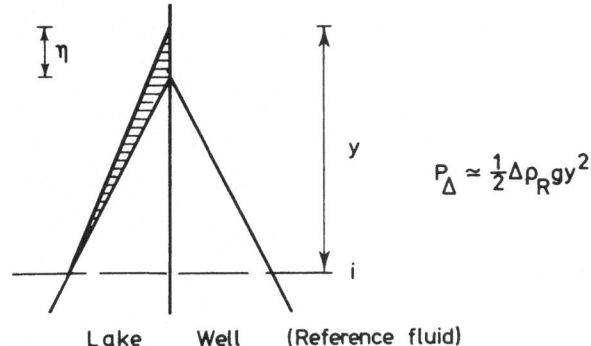

Fig. 2.4 *The pressure distribution in the lake and the connected well. The hatched area is the excess pressure (highly distorted scale).*

Secondly, we want to illustrate the exaggeration effect on the interfacial slope.

As $\eta = \Delta y$ is a growing η in the x-direction associated with a growing depth y, i.e. the slope of the water surface, s, and of the interface, i, is deflected in opposite directions. Hence, taking the surface slope I_s as positive, we obtain, by definition

$$I_s = \frac{\partial \eta}{\partial x} = \Delta \frac{\partial y}{\partial x} \tag{2.12}$$

and for the interfacial slope I_i

$$I_i = - \frac{\partial(y-\eta)}{\partial x} \simeq - \frac{\partial y}{\partial x} \tag{2.13}$$

(as $\eta = \Delta y \ll y$)

or

$$I_i \simeq - \frac{I_s}{\Delta} \tag{2.14}$$

which illustrates the strong exaggeration on the interfacial slope, as Δ is very small in nature, often 10^{-2} to 10^{-3}.

The potential energy

Now, let us turn to the second subject of this chapter, the potential energy (POT). We cut a horizontal slice of thickness dz of the stratified body of water shown in Fig. 2.5. The contribution from this slice to the potential energy of the whole water body corresponds to the work to be done against gravity to raise it to the level of reference (here arbitrarily chosen as the water surface)

$$d\ POT = -\ (A\rho dz)gz \qquad (2.15)$$

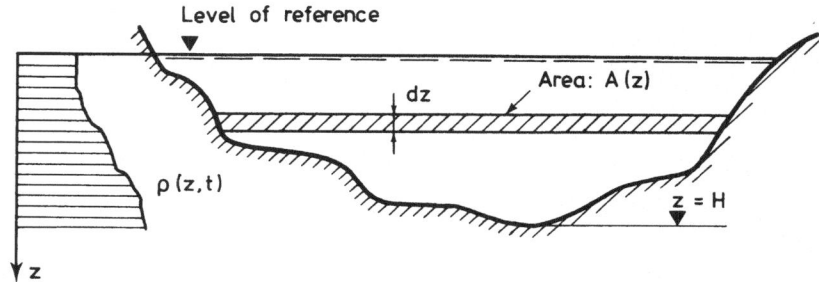

Fig. 2.5 *Density stratified lake with varying cross-sectional area.*

Therefore, the total potential energy contained in the lake is

$$POT = -\int_0^H \rho(z,t)g\ A(z)z\ dz \qquad (2.16)$$

(where t is time). Similar to the pressure calculations, it is convenient to operate in a reduced system where the potential energy reads

$$POT_\Delta = -\int_0^H (\rho(z,t)-\rho_R)g\ A(z)\ z\ dz \qquad (2.17)$$

or, if we introduce the dimensionless reduced mass Δ

$$POT_\Delta = -\int_0^H \Delta(z,t)\rho_R\ g\ A(z)\ z\ dz \qquad (2.18)$$

12

In the following we omit the index Δ in POT, as confusion seems unlikely to occur.

Example 2.3

We consider 1 m^2 of the two-layer water column in Fig. 2.6, for which the change in potential energy due to a total mixing will be calculated.

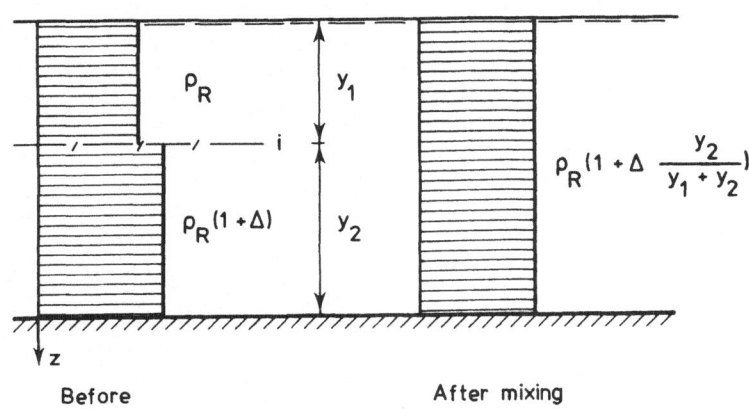

Before   After mixing

Fig. 2.6 Total mixing of an initially two-layered water column.

At first the density of the mixed water is determined (constancy of mass)

$$\rho_R(y_1 + y_2 + \Delta y_2) = \rho_R(1 + \Delta_M)(y_1 + y_2) \qquad (2.19)$$

which yields the dimensionless reduced mass of the mixed fluid

$$\Delta_M = \Delta \frac{y_2}{y_1 + y_2} \qquad (2.20)$$

Then we calculate the reduced POT of the two-layer column

$$POT_{Before} = - \int_0^{y_1+y_2} \Delta(z)\rho_R \, g \, z \, dz =$$

$$- \Delta\rho_R g \left[\frac{1}{2}(y_1+y_2)^2 - \frac{1}{2} y_1^2 \right] \qquad (2.21)$$

13

Similarly, after the mixing, we get

$$POT_{After} = - \Delta_M \, \rho_R g \left[\frac{1}{2}(y_1+y_2)^2 \right] \qquad (2.22)$$

The change in POT, which is equal to the work needed for the mixing process or, in other words, the potential energy gained, is

$$\Delta POT = POT_{After} - POT_{Before} = \frac{1}{2}\Delta\rho_R g y_1 y_2 \qquad (2.23)$$

This result can be achieved directly, if we interpret the result as the work done by raising the excess mass, $\Delta\rho_R y_2$, the center of gravity distance $1/2 \, y_1$.

Exercise 2.1

Evaluate the change in POT associated with a total mixing of a water column with initially linearly distributed density $\rho_R (1 + \Delta\frac{z}{H})$.

Result: $\Delta POT = \frac{1}{12} \Delta\rho_R g H^2$

Show that a center of gravity consideration leads to the same result.

Up to now only stable stratifications have been considered, where a mixing is associated with a gain in potential energy, i.e. caused by an external energy supply. Some processes taking place in nature cause a release of potential energy, namely where an unstable stratification is formed, for instance by evaporation in the sea or by cooling/heating in a lake with an initial temperature of $> 4°C/ < 4°C$, respectively. In Fig 2.7 the evaporation process in the Mediterranean is illustrated. As the water evaporated is fresh, the evaporation process releases brine (a heavy salt solution) at the surface, and this brine has

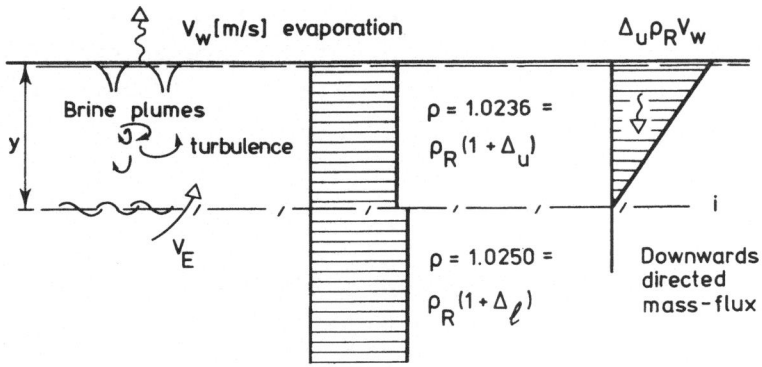

*Fig. 2.7 Schematic illustration of the evaporation pro-
cess in the mediterranean.*

a higher density than the surrounding sea-water (in the example
the rate of brine-release is $\Delta_u \rho_R V_w = 23.6$ (kg salt/m^3) $\times V_w$,
where V_w is the rate of evaporation).

The associated buoyancy flux creates a high level of turbulence in the upper layer y of homogeneous water, which means
that the salt released is evenly distributed over the depth y -
or, accordingly, that the salt - or mass - flux is linearly distributed with a maximum value $\Delta_u \rho_R V_w$ [kg/m^2/sec] at the free surface. Hence, the released potential energy, or the potential
energy lost per unit time, amounts to

$$POT = \frac{1}{2} (\Delta_u \, \rho_R \, V_w) g \; y \qquad\qquad (2.24)$$

Example 2.4

Let us illustrate the mixing-efficiency concept for the
evaporation process shown in Fig. 2.7. The falling brine plumes
create a high level turbulent upper layer, which erodes the interface at a rate V_E [m/s]. The entrained denser water from below gains potential energy by being lifted (by turbulent diffusion) the distance $\frac{1}{2}$ y on the average, yielding a potential energy per unit time which amounts to

$$POT_{Gain} = \frac{1}{2} (\Delta_\ell - \Delta_u) \rho_R \, V_E \; g \; y \qquad\qquad (2.25)$$

(where index ℓ = lower and u = upper layer).

The process described - called entrainment by free penetrative convection - is empirically known to have an efficiency of about 18%, and hence the rate of erosion, V_E, is determined by the relation

$$POT_{Gain} = 0.18 \ POT_{Lost} \tag{2.26}$$

where POT_{Lost} is given by equation (2.24). Inserting the two expressions evaluated for POT in equation (2.26) we obtain the entrainment velocity

$$V_E = \frac{\Delta_u}{\Delta_\ell - \Delta_u} \ V_w \times 0.18 \tag{2.27}$$

Note that the salt released from above as well as the salt entrained from below increase the mass of the upper layer Δ_u (and hence V_E) as time goes by, which means that the interface - erosion - process is an accelerating process.

Exercise 2.2

We consider 1 m^2 of the water column in Fig. 2.6, example 2.3, exposed to a wind generated shear stress $\tau = 2.6 \times 10^{-3} \ \rho_a W^2$ ($\rho_a \simeq 1/800 \ \rho_R$ = density of air; W = 10 m/s = wind velocity). The surface generated velocity is $U_s \simeq 1/30$ W, and hence the total mixing energy-input per unit time amounts to τU_s. We take the depth y_1 = 10 m and y_2 = 20 m, respectively, and the dimensionless reduced mass $\Delta = 10^{-3}$.

Question: For how long time must the wind blow in order to mix the whole water column, provided the efficiency of the mixing process is, say 5% ?

Answer: 1.9×10^5 sec $\simeq$ 2.2 days.

Later on we shall outline more general methods for evaluating the mixing conditions in stratified fluids. In these methods the change (gain) in potential energy quite often plays the most crucial role.

3. THE MOTION OF NON-MISCIBLE STABLY STRATIFIED FLUIDS

The most common stratified flows in nature are associated
with mixing. On the other hand, flow situations exist where
either no mixing takes place (oil/water) or where the actual
mixing has no appreciable influence on the dynamics of the flow,
as for instance the stationary salt water wedge (treated below).
Therefore, we have an ample motivation to discuss the motion
of non-miscible stratified fluids, in as much as it leaves us
the possibility to get familiar with the special effects of stratified flows in a relatively simply manner. Generally speaking,
the non-miscible stratified flows behave like ordinary open
channel flows in a reduced acceleration of gravity field $g' = \Delta g$,
confer Appendix.

Caution! This general concept does not always apply for
the miscible stratified fluids.

In order to illustrate the resemblance to the hydraulics
of open channel flow, we start to look at the flow of a dense
bottom current (under-flow). Then we calculate the stationary
salt water wedge (over-flow), and finally a number of oil/water
flow situations, including wind set-up, are treated.

3.1 Dense bottom currents

We consider a one-dimensional flow of a heavy fluid along
the bottom of a deep basin with a light fluid of density ρ_R, see
Fig. 3.1.1. (Examples of natural occurrence of this type of flow
in Chapter 8). The effect on the flow of a local hump can be investigated by treating the flow as inviscid, i.e. the Bernouilli's equation applies.

We choose a streamline along the bottom from point A to
point B:

$$\rho_R \, g \, D + \Delta\rho_R \, g \, y_A + \frac{1}{2} \, \rho_R \, (1 + \Delta) \, V_A^2 =$$

$$\rho_R \, g \, D + \Delta\rho_R \, g \, y_B + \frac{1}{2} \, \rho_R \, (1 + \Delta) \, V_B^2 = \qquad (3.1.1)$$

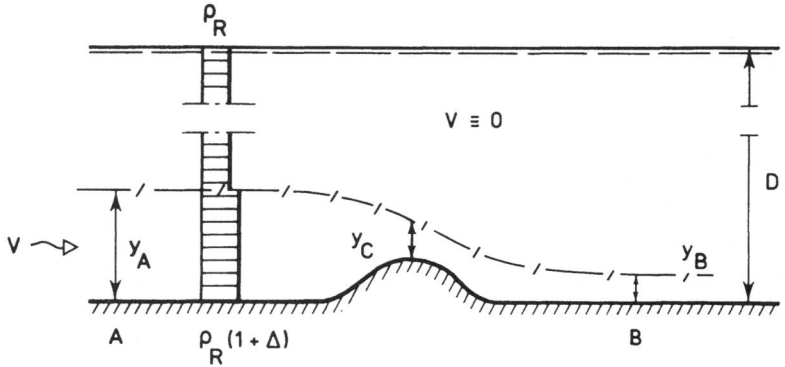

Fig. 3.1.1 Dense bottom current across a sill (potential flow).

Notice that the pressure term $p_R = \rho_R \, g \, D$, originating from the reference fluid, is present in both points, and hence could have been omitted. Furthermore, we may neglect the difference between ρ_R and $\rho_R(1 + \Delta)$ in the inertia-terms (the Boussinesq approximation), but of course not in the pressure terms.

With these approximations we get

$$y_A + \frac{V_A^2}{2\Delta g} = y_B + \frac{V_B^2}{2\Delta g} \qquad (3.1.2 \text{ a})$$

or

$$y_A + \frac{V_A^2}{2g'} = y_B + \frac{V_B^2}{2g'} \qquad (3.1.2 \text{ b})$$

in analogy to open channel flow, except for g, which has been replaced by $g' = \Delta \, g$.

Accordingly, the specific energy of the dense bottom current reads

$$H_0 = y + \frac{V^2}{2g'} \qquad (3.1.3)$$

which has a minimum (for constant discharge q per unit width) when

$$\frac{dH_0}{dy} = 1 - \frac{q^2}{g'y^3} = 0 \qquad (3.1.4)$$

18

which is the well-known condition for the critical depth

$$y_c = \sqrt[3]{\frac{q^2}{g'}} \qquad (3.1.5)$$

It therefore becomes natural to introduce the densimetric Froude's number

$$\mathbb{F}_\Delta = \frac{V}{\sqrt{g'y}} = \frac{V}{\sqrt{\Delta gy}} \qquad (3.1.6)$$

Finally, if we now return to real fluids, it is evident that all the well-known phenomena and relations applicable in open channel flow can be directly transferred to dense bottom currents of immiscible fluids, provided we treat the current as if the acceleration of gravity were reduced to Δg, i.e. takes place in a "slow-motion" fashion. This applies for back-water curves, hydraulic jumps etc. In part II of the present lecture notes, we shall learn how the mixing modifies the results achieved for non-miscible stratified flows.

3.2   The stationary salt water wedge

When the light fresh water from a river reaches the more dense salt water in the sea, the light water will ride over the salt water wedge which has intruded the river due to its higher density, confer with Fig. 3.2.1.

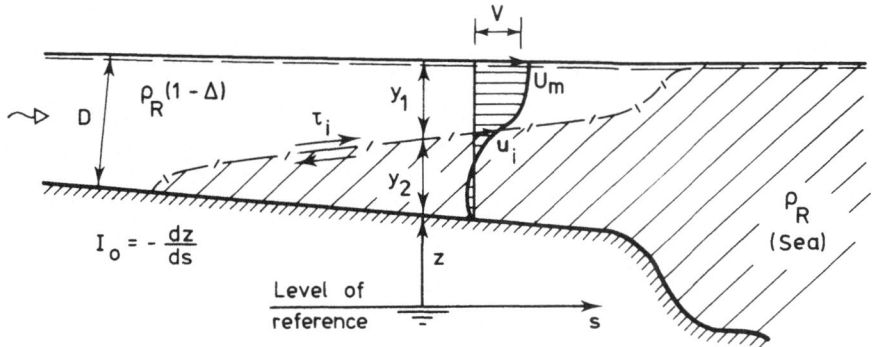

Fig. 3.2.1 *The stationary salt water wedge (highly distorted scale).*

We consider a salt water wedge of unit width in which the discharge q is constant (i.e. no mixing), and where no wind, tides etc. are present. Under these circumstances we have a net outwards directed discharge in the upper fresh layer, and a circulation (with no net flow) in the lower salty layer, created by the interfacial shear stress

$$\tau_i/\rho = f_i/2 \ (U_m - u_i)^2 = f/2 \ V^2 \tag{3.2.1}$$

Here the friction coefficient for the interface $f_i/2$ is associated with the maximum minus minimum velocity in the flow (see Chapter 7). For convenience we have related τ_i to the depth average velocity V in the present example, which means that the associated friction coefficient $f/2$ only amounts to say 1/3 to 1/4 of the value known from open channel flow. The mixing across the interface in a stationary salt water wedge does not influence the dynamics of the salt water wedge, and will therefore be neglected.

The shear stress transferred to the interface is balanced by a longitudinal pressure gradient in the lower fluid (negligible momentum).

$$\tau_i = \rho_R \ g \ y_2 \ \frac{\partial p_\ell/\rho_R g}{\partial s} =$$

$$\rho_R \ g \ y_2 \left[(1-\Delta) \ \frac{dy_1}{ds} + \frac{dy_2}{ds} + \frac{dz}{ds} \right] \tag{3.2.2}$$

(where p_ℓ is the pressure in the lower layer).

For the upper layer flow, we may apply the ordinary depth integrated energy equation, which reads (the velocity distribution coefficient $\alpha \simeq 1$)

$$I = - \frac{d}{ds} \left[z + y_2 + y_1 + \frac{V^2}{2g} \right] =$$

$$I_0 - \frac{dy_2}{ds} - \left(1 - \mathbb{F}_1{}^2 \right) \frac{dy_1}{ds} \tag{3.2.3}$$

which is the well-known differential equation for open channel flow. Notice that contrary to the underflow considered in the preceeding example, the salt water wedge is an overflow with a free surface and hence exposed to the full acceleration of gravity.

If we - instead of the energy equation - had used the depth integrated momentum equation, an expression similar to equation (3.2.3) had appeared, provided we had replaced the energy-gradient I with the interfacial shear stress τ_i in the following well-known way

$$\tau_i = \rho_R (1-\Delta) g \, y_1 \, I \simeq \rho_R \, g \, y_1 \, I \qquad (3.2.4)$$

By combining equations (3.2.2) and (3.2.3), the thickness of the salt water wedge can be partly eliminated

$$\frac{\tau_i}{\rho_R g y_2} - (1-\Delta) \frac{dy_1}{ds} + I_0 = I_0 - I - \left(1 - \mathbb{F}_1{}^2\right) \frac{dy_1}{ds} \qquad (3.2.5)$$

which by introducing equation (3.2.4) is further reduced to

$$I \left(\frac{y_1}{y_2} + 1\right) = - \left(\Delta - \mathbb{F}_1{}^2\right) \frac{dy_1}{ds} \qquad (3.2.6)$$

In order to proceed further, we introduce the total depth $D = y_1 + y_2$ (which for convenience is taken as a constant in the following) and the friction coefficient by

$$\frac{I}{\Delta} = \frac{f/2 \ V^2}{\Delta \ g \ y_1} = f/2 \ \mathbb{F}_{\Delta,0}{}^2 \left(\frac{D}{y_1}\right)^3 \qquad (3.2.7)$$

where the (known) upstream densimetric Froude number squared

$$\mathbb{F}_{\Delta,0}{}^2 = \frac{q^2}{\Delta \ g \ D^3} \qquad (3.2.8)$$

has been used as a dimensionless parameter.

Hence, the governing differential equation for the salt water wedge - or rather for the flow above the wedge - yields

$$\frac{d\eta}{d\xi} = \frac{-f/2 \ \mathbb{F}_{\Delta,0}{}^2}{(1-\eta)(\eta^3 - \mathbb{F}_{\Delta,0}{}^2)} \qquad (3.2.9)$$

(where $\eta = y_1/D$ and $\xi = s/D$).

At $\eta = 1$ (at the toe of the wedge) and at $\eta = \mathbb{F}_{\Delta,0}{}^{2/3}$ (at the heal of the wedge, $\xi = 0$) $d\eta/d\xi$ is infinite, corresponding to an interface perpendicular to the bottom. Of course, this makes no physical sense, but is a well-known consequence of neglecting the curvature of the streamlines.

From equation (3.2.9) it is evident that the salt water wedge only exists if

$$\mathbb{F}_{\Delta,0}{}^2 = \frac{q^2}{\Delta\ g\ D^3} < 1 \qquad (3.2.10)$$

because η has to be smaller than one, and furthermore $d\eta/d\xi$ has to be negative. As the order of magnitude of Δ in common cases is $\Delta \sim 0.028$, the condition corresponds to an ordinary Froude number of $\mathbb{F}_0 = q/\sqrt{gD^3}$ of say 0.17. Hence, when the river Froude number is greater than say 0.17, no harmful salt-intrusion takes place in that river.

Example 3.2.1
=============

We consider a unit width of an estuary (without Coriolis effects), in which the densimetric Froude number of the river is $\mathbb{F}_{\Delta,0} = 0.2$ (<1), and in which the interfacial friction factor (related to V) is $f/2 = 6 \times 10^{-4}$. The dimensionless critical depth at the river mouth ($\xi = 0$) is

$$\eta_0 = \mathbb{F}_{\Delta,0}{}^{2/3} = 0.342 \qquad (3.2.11)$$

The differential equation (3.2.9) may readily be solved to yield

$$\int_{\eta}^{\eta_0} \left(\eta^3 - \eta^4 - \mathbb{F}_{\Delta,0}{}^2 + \mathbb{F}_{\Delta,0}{}^2\ \eta \right) d\eta = -\int_{\xi}^{0} f/2\ \mathbb{F}_{\Delta,0}{}^2\ d\xi \qquad (3.2.12)$$

or

$$1/4\ \eta^4 - 1/5\ \eta^5 - \mathbb{F}_{\Delta,0}{}^2\ \eta + 1/2\ \mathbb{F}_{\Delta,0}{}^2\ \eta^2 \ \bigg|_{\eta}^{\eta_0} = f/2\ \mathbb{F}_{\Delta,0}{}^2\ \xi$$

$$(3.2.13)$$

shown in Fig. 3.2.2.

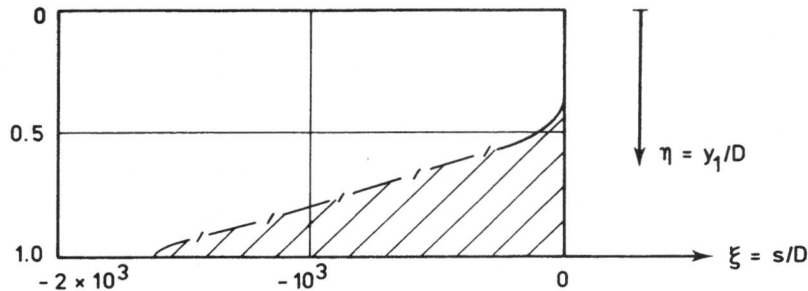

Fig. 3.2.2 The dimensionless salt water wedge
$(I\!\!F_{\Delta,0} = 0,2;\ f/2 = 6 \times 10^{-4})$.

If we take a water depth of D = 10 m, the extension of the
salt water wedge is seen to be approximately 16 kilometer. The
order of magnitude of the energy gradient can be estimated by
using equation (3.2.7). If we introduce $\Delta = 0.028$ and an average value of $D/y_1 \approx 1.5$, we obtain $I_{average} \sim 2 \times 10^{-6}$, which is
a considerably low value (caused by the low interfacial friction).
From Fig. 3.2.2 the slope of the interface can - on average -
be estimated to $I_i \sim 3 \times 10^{-4}$, which is not equal to the surface
slope divided by Δ (as in Eq. (2.14)), because the lower layer
has a non-zero pressure gradient.

———————

Caution: Laboratory generated salt water wedges have a
length to depth ratio which is considerably smaller than the
length to depth ratio in nature. This implies that the interfacial shear stress in the laboratory over a relatively great
length is dominated by the growing boundary layer, starting at
the toe of the wedge. As the friction coefficient for a growing
boundary layer is about 10 times as great as for a fully developed flow, a simple geometric scaling from laboratory to field
salt water wedges is unjustifiable.

3.3 Containment of oil spill in rivers and harbours

The heavy traffic of oil-tankers on the sea and in the
large rivers involves a great potential risk for oil pollution

of our aquatic environments. Not only the unique accidents blown
up in the newspapers, but also the daily "accidential" oil spill
are great hazards to the aquadic life.

One of the more effective tools for limiting an inevitable
hazard to life from an oil spill is to contain the oil within a
restricted area by the use of floating booms, and in this way
form a thickness of the oil, sufficient for a following pumping
of the oil. The effectivity of a boom may be limited by the local wind currents and waves, and hence the method is most useful in sheltered areas like for instance harbours and rivers.

In the present chapter we shall formulate the conditions
for containment of an oil spill by floating booms under various
flow and wind conditions.

3.3.1 Containment of oil in a current

In Fig. 3.3.1 a stationary situation is shown, where all
the spilled oil has been contained by a floating boom laid out
across the river.

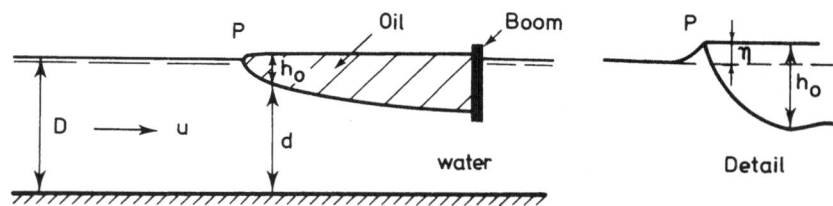

Fig. 3.3.1 Containment of an oil spill in a river.

Let the river characteristics be the depth D, the depth
average velocity u, the density $\rho (\simeq 10^3$ kg/m^3). The density of
the oil is $\rho_0 = \rho(1-\Delta)$, where Δ is up to say 0.1.

The stagnation pressure at P creates a "head" with a super
elevation η of the water level, which can be determined by Bernouilli's equation for a surface streamline

$$\eta = \frac{u^2}{2g}$$

(3.3.1)

or in dimensionless form

24

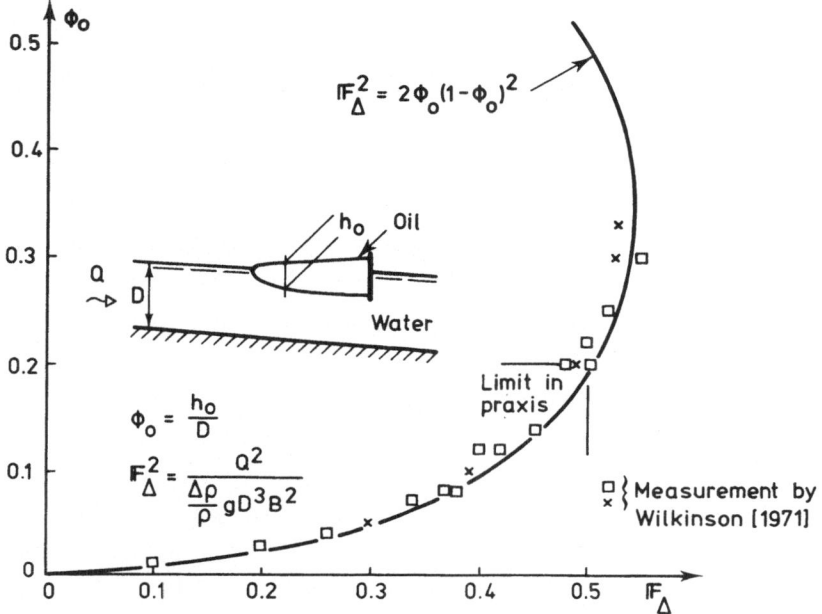

Fig. 3.3.2   *The dimensionless start-thickness of the oil pool as a function of the densimetric Froude number.*

$$\frac{\eta}{\Delta D} = \frac{1}{2} \, \mathbb{F}_\Delta^{\,2} \qquad\qquad (3.3.2)$$

(where $\mathbb{F}_\Delta^{\,2} = u^2/\Delta gD$).

The convergence of the streamlines below the oil pool creates a pressure drop without any significant head loss. The Bernouilli's equation states, this time for a streamline along the bottom.

$$D + \frac{u^2}{2g} = d + \frac{\rho_0}{\rho} \, h_0 + \frac{u^2}{2g} \left(\frac{D}{d}\right)^2 \qquad\qquad (3.3.3)$$

where h_0 is the oil thickness at the head. The left hand side of equation (3.3.3) is equal to $D + \eta$ (equation 3.3.1) or according to Fig. 3.3.1 equal to $(d + h_0)$ as well. Therefore Eq. (3.3.3) can be reduced to yield

$$\Delta h_0 = \frac{u^2}{2g} \left(\frac{D}{d}\right)^2 \simeq \frac{u^2}{2g} \left(\frac{D}{D-h_0}\right)^2 \qquad\qquad (3.3.4)$$

or in dimensionless form

$$\mathbb{F}_\Delta{}^2 = 2\phi_0 \; (1 - \phi_0)^2 \tag{3.3.5}$$

(where $\phi_0 = h_0/D$). Eq. (3.3.5), which expresses the upstream oil-thickness h_0 as a function of the river densimetric Froude number, is plotted in Fig. 3.2.2, where measurements performed by D. Wilkinson [1971] are shown as well. From Fig. 3.2.2 we obtain the condition for oil containment in practise

$$\mathbb{F}_\Delta < \; \sim 0.5 \text{ or } \phi \leq \sim 0.2 \tag{3.3.6}$$

Example 3.3.1

In a specific river where a risk for an oil spill is pre-sent, the yearly variation of the river densimetric Froude number at a certain station is illustrated in Fig. 3.3.3.

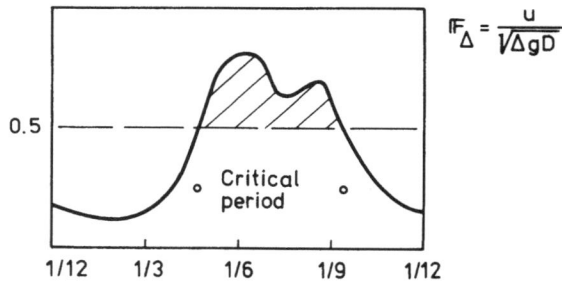

Fig. 3.3.3   The yearly variation of the densimetric Froude number for a specific location in a river.

The diagram shows clearly that during the period 15/4 to 15/9 it is impossible to contain any oil spill at that location by use of booms. Similar diagrams for other stations in the actual river would together form a master emergency plan showing where and when a certain oil type may be contained by booms across the river.

Example 3.3.2

If the water depth becomes "infinite", ϕ_0 goes to zero, which means that the factor $(1 - \phi_0)^2$ may be omitted in equation (3.3.5), i.e. the condition for oil containment simplifies to

$$\mathbb{F}_\Delta^2 \simeq 2\phi_0 \tag{3.3.7}$$

applicable for the spreading of oil in the sea. Denoting the spreading velocity by c, Eq. (3.3.7) yields

$$c = \sqrt{2\Delta g h_0} \tag{3.3.8}$$

an expression verified by experiments. As the surface tension has been neglected, Eq. (3.3.8) cannot be used for calculations of oil-films, for which reference is made to Hoult [1972].

Eq. (3.3.8) may be used to calculate the relation between the submerged depth of a certain boom and its towing speed, if oil flow below the boom is to be avoided.

Example 3.3.3

In the other limit, we may solve the classical "lock-exchange" problem by using Eq. (3.3.5). Fig. 3.3.4 illustrates the transformation needed.

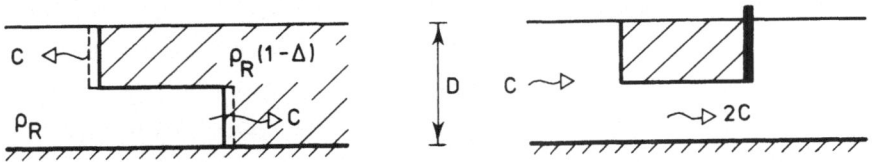

Fig. 3.3.4   The non-steady "lock-exchange" problem transformed to a stationary flow in a moving frame of reference.

In the present case $\phi_0 = 0.5$ and hence $\mathbb{F}_\Delta^2 = 1/4$ or

$$c = 0.5 \sqrt{\Delta g D} \tag{3.3.9}$$

in accordance with experiments.

After the problem of the condition for oil containment by booms has been elucidated, the question arises: what is the amount of oil collectable? This amount depends on the downstream variation of the oil thickness, which can be determined by using the momentum equation for the oil and the water flow, respectively. The interfacial shear stress τ_i only creates a weak circulation in the oil pool, and hence τ_i is balanced by a set-up of the oil surface, see Fig. 3.3.5,

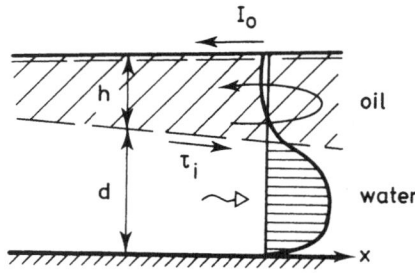

Fig. 3.3.5 *Longitudinal section of a river in which oil has been contained by a floating boom (located downstream).*

$$\tau_i = \rho_0 \ g \ h \ I_0 \simeq \rho \ g \ h \ I_0 \qquad (3.3.10)$$

The momentum equation for the water flow is rather time consuming to solve in the general case (Bo Pedersen [1974b]), so we shall restrict ourselves to a very simple case, namely where the river width B is much greater than the depth d, which in turn is much greater than the oil-thickness h. If we further presume that the river (ordinary) Froude number is small, we may neglect the longitudinal pressure gradient in the water, which means that the superelevation of the oil-surface is $\eta = \Delta h$ (see equation 2.11). Inserted in Eq. (3.3.10) the following simple differentialequation for the oil-thickness h appears

$$f_i/2 \ \rho \ u^2 = \rho \ g \ h \ \frac{d(\Delta h)}{dx} \qquad (3.3.11)$$

(where $\tau_i/\rho = f_i/2 \ u^2$ has been introduced).
    The solution to Eq. (3.3.11) is

$$h^2 - h_0^2 = f_i \ \frac{u^2}{\Delta g} \ x \qquad (3.3.12)$$

where the upstream thickness h_0 can be determined by Eq. (3.3.5) ($\phi_0 \ll 1$).

Rewritten in dimensionless form, the solution reads

$$\{\frac{\Delta gh}{u^2}\}^2 = 0.25 + f_i \frac{\Delta gx}{u^2} \tag{3.3.13}$$

In the present case, the friction coefficient is rather high, namely corresponding to a growing boundary layer in the underflowing water. Therefore we put $f_i/2 \simeq 5 \times 10^{-3}$ and obtain

$$\frac{\Delta gh}{u^2} = 0.1 \sqrt{\frac{\Delta gx}{u^2} + 25} \tag{3.3.14}$$

which has been compared with experiments reported on by Cross and Hoult [1971], Fig. 3.3.6

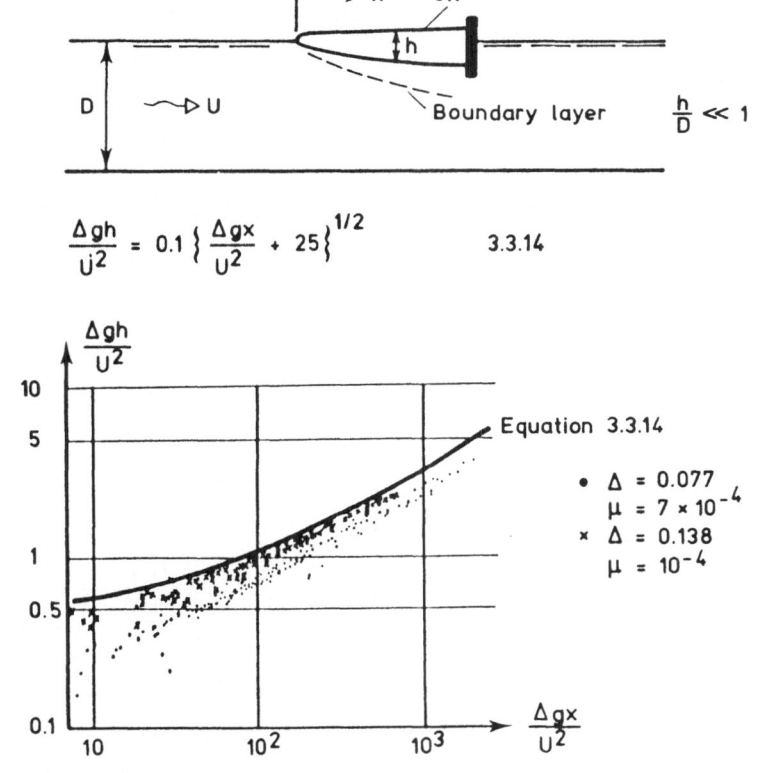

Fig. 3.3.6   *Experiments on containment of oil by a boom laid across a deep river.*

The total amount of contained oil per unit width V_{oil} can be calculated by performing an integration

$$V_{oil} = \int_0^x h \ dx =$$

$$0.1 \ \frac{2}{3}\left(\frac{u^2}{\Delta g}\right)^2 \left[\left(\frac{\Delta g x}{u^2} + 25\right)^{3/2} - 25^{3/2}\right] \qquad (3.3.15)$$

If we introduce the oil-thickness at the downstream end h_b (at the boom) and neglect the last term on the right hand side of Eq. (3.3.15) (normally an insignificant contribution to V_{oil}), we finally get (per unit width)

$$\frac{V_{oil}}{h_b^{\ 2}} \simeq 67 \ \{\frac{\Delta g h_b}{u^2}\} \qquad (3.3.16)$$

Exercise 3.3.1

What is the maximum amount of oil with $\Delta = 0.05$ which can be contained by a boom submerged $h_b = 0.4m$ in a river with discharge $Q = 10 \ m^3/s$, depth $D = 2 \ m$ and width $B = 15 \ m$?

Answer: 284 m^3.

What is the length of the oil pool ?

Answer: 63 m.

_____

3.2.2 Wind set-up in oil on water

When the wind blows along the river, it will either reduce or increase the amount of collectable oil when blowing in the downstream or upstream direction, respectively. The influence of the wind can be taken into account by adding the imposed wind generated shear stress on the oil-surface in the preceeding equations of motion. We take a simple case to illustrate the wind effects, Fig. 3.3.7

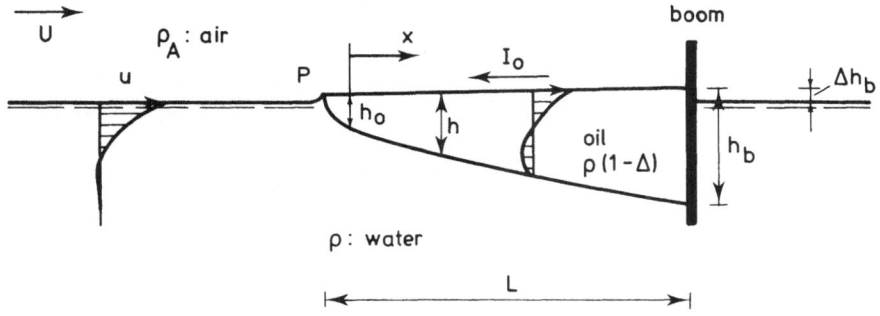

Fig. 3.3.7 Containment of oil on water in a sea exposed
to the wind.

The wind induced surface velocity in the water may be estimated by assuming equal shear stress in the two media at the
interface

$$\rho u^2 = \rho_A U^2 \qquad (3.3.17)$$

The upstream start thickness is determined by the stagnation pressure in P, confer with Eq. (3.3.1) and (2.11),

$$h_0 = \frac{u^2}{2g\Delta} = \frac{\rho_A}{\rho} \frac{U^2}{2g\Delta} \qquad (3.3.18)$$

The momentum equation for the oil pool simply states that
the imposed wind shear stress is balanced by the pressure gradient in the oil, as the induced circulation is very weak

$$\tau_w = \frac{d}{dx} \left(\frac{1}{2}\Delta\rho g h^2 \right) \qquad (3.3.19)$$

which may readily be integrated to yield

$$\tau_w L = \frac{1}{2}\Delta\rho g \{h_b{}^2 - h_0{}^2\} \qquad (3.3.20)$$

(where L is the length of the oil pool).

If we take an average value for the friction coefficient
($\tau_w = 2.3 \times 10^{-3} \rho_A U^2$) and for the ratio $(\rho_A/\rho) \simeq 1.15 \times 10^{-3}$,
Eq. (3.3.20) may be rearranged to yield

$$\frac{h_b}{L} \simeq 2.4 \times 10^{-3} \frac{U}{\sqrt{\Delta gL}} \{1 + 0.065 \frac{U^2}{\Delta gL}\}^{1/2} \tag{3.3.21}$$

which is drawn in Fig. 3.3.8 and compared with experiments reported on by Sorensen and Spencer [1971].

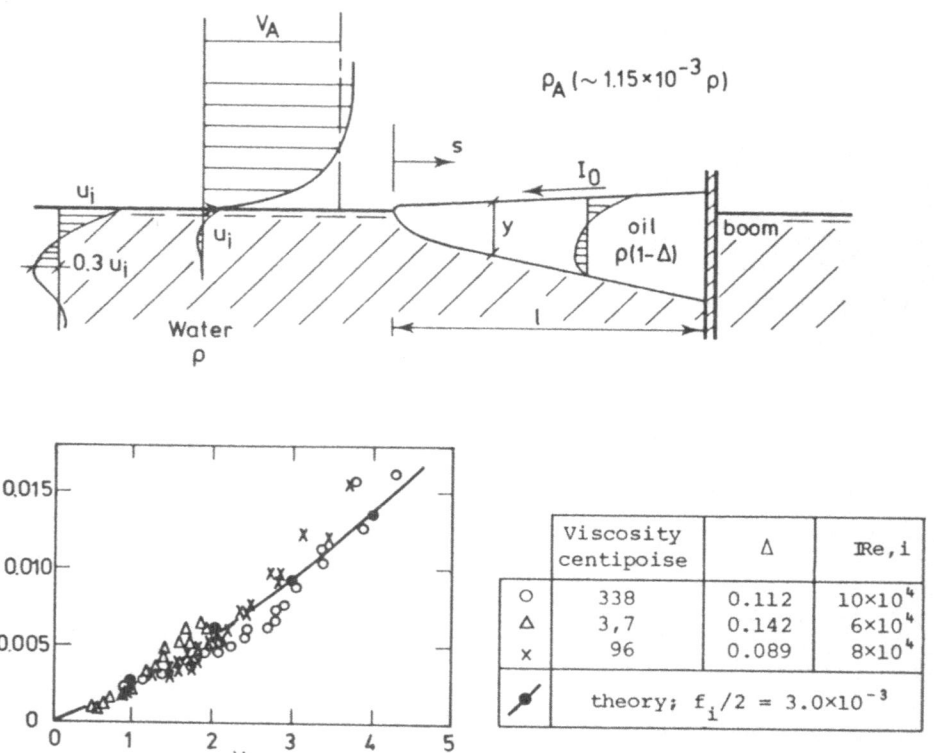

Fig. 3.3.8 *Wind-piling of oil on water.*

Accordingly, we may evaluate the total amount of contained oil per unit boom-width by integrating Eq. (3.3.21)

$$V_{oil} = \left\{ \frac{11.6 \times 10^4}{\left(\frac{U^2}{\Delta gh_b}\right)} - 26.5 \times 10^{-6} \left(\frac{U^2}{\Delta gh_b}\right)^2 \right\} h_b{}^2 \tag{3.3.22}$$

32

Exercise 3.3.2

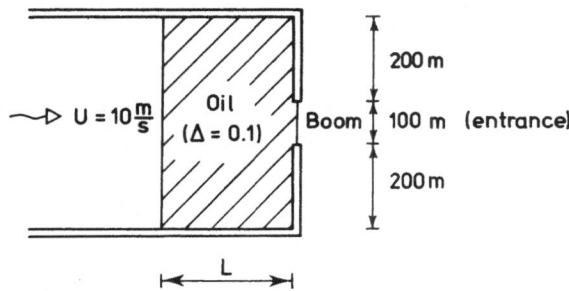

Fig. 3.3.9   Sketch of a harbour with floating submerged booms laid across the entrance.

In the above sketched harbour 0.4 m submerged booms are available for keeping spilled oil withing the harbour area. What is the maximum length L of the oil pool for the wind condition shown?

Answer: L = 265 m.

What is the maximum volume of oil which can be kept within the harbour?

Answer: 36×10^3 m^3.

PART II

MISCIBLE STRATIFIED FLOWS

- CHAPTERS 4 - 12 INCL. -

4. THE EQUATIONS OF CONTINUITY AND MOTION FOR MISCIBLE STRATIFIED FLOWS

The motion of real fluids is associated with an energy loss, i.e. - as far as turbulent flow is concerned - with a production of turbulent kinetic energy, which in a homogeneous flow is dissipated into heat. In a stratified flow, part of this turbulent kinetic energy is used to mix the flowing water with the ambient water. Let us briefly illustrate this mixing process in a dense bottom current, see Fig. 4.1

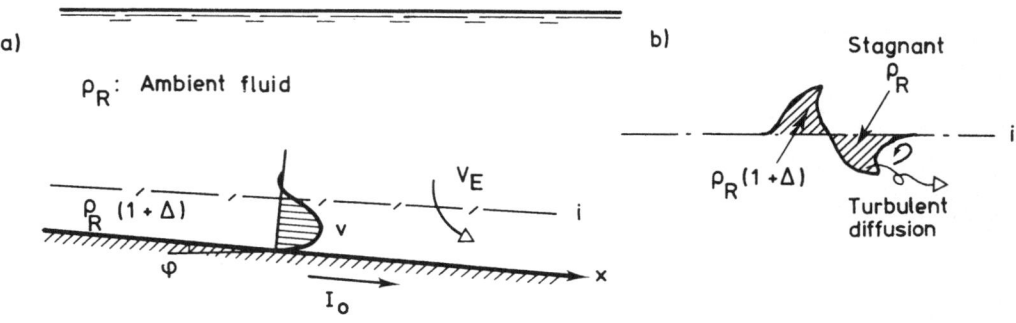

Fig. 4.1   a) Two layer stratified flow (dense bottom
              current).
           b) Deformation of the interface i.

Any deformation of the interface (Fig. 4.1b) will produce a restoring body force, forcing the elements back to their respective regions. During the excursion, the light ambient water has been surrounded by a turbulent fluid, and hence part of it has been carried away by turbulent diffusion, making it an integrated part of the dense bottom current. Entrainment has taken place. On the contrary, the dense fluid element has not experienced any turbulent diffusion process, and therefore is returned to the flow without any loss of mass. The entrainment process is a one-way transport process from the ambient, stagnant water to the flowing turbulent water. Entrainment is treated in more detail in chapter 6.

The implications of the entrainment process are an increased volume flux, which has to be taken into account in the continuity equation for volume; a changed reduced mass Δ (decreasing in the flow direction), which has to be incalculated into the mass-continuity equation and in the pressure terms appearing in the equations of motion. Besides a need for modifying the basic equations there is a need for relating the rate of entrainment V_E to the bulk parameters of the flow. To this end we shall outline and discuss the energy equation for the turbulence.

4.1 The continuity equations

The equation for conservation of mass in turbulent flow (neglecting molecular diffusion) is

$$\frac{d\rho}{dt} + \rho \frac{\partial v_i}{\partial x_i} = 0 \qquad (4.1.1)$$

where d/dt means total differentiation (following the motion), i.e.

$$\frac{d\rho}{dt} = \frac{\partial \rho}{\partial t} + v_i \frac{\partial \rho}{\partial x_i} \qquad (4.1.2)$$

(where summations have to be performed when subscript is repeated).

Dealing with homogeneous fluids under moderate pressure conditions, we assume the fluid to be incompressible, i.e. $d\rho/dt = 0$. This is not always the case in inhomogeneous fluids, which will be illustrated in the following example.

Example 4.1.1

Suppose we have a stationary situation, where two different streams A and B form a third stream C, which becomes a perfect mixture of A and B. Under these circumstances, the continuity equation reads

$$\rho \frac{\partial v_i}{\partial x_i} + v_i \frac{\partial \rho}{\partial x_i} = \frac{\partial}{\partial x_i} (\rho v_i) = 0 \qquad (4.1.3)$$

The characteristics of A and B may be (see Fig. 4.1.1)

$$T_A = 0^{\circ}C \qquad S_A = 11.3\ ^0/_{00} \qquad \sigma_A = 9.0\ kg/m^3$$

$$T_B = 16^{\circ}C \qquad S_B = 13.1\ ^0/_{00} \qquad \sigma_B = 9.0\ kg/m^3$$

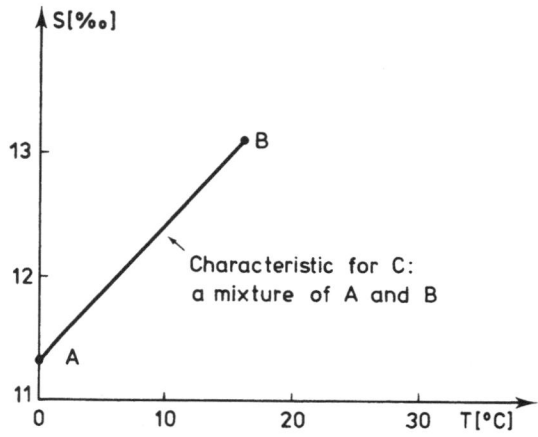

Fig. 4.1.1 T-S diagram for mixing of fluid A and B.

i.e. the same density. A T-S diagram (Fig. 4.1.1) is within
oceanography commonly used to identify the origin of a volume
of mixed water. If for the sake of simplicity A and B contri-
bute at an equal rate to C, we get

$$T_C = 8 \ ^oC \qquad\qquad S_C = 12.2 \ ^0/_{00} \qquad\qquad \sigma_C = 9.5 \ kg/m^3$$

(see Fig. 1, appendix, with respect to σ_C). The mixture has a
density $\sigma_C = 9.5 \ kg/m^3$, inspite of the fact that it is a mixture of two fluids, both with a density of $\sigma_A = \sigma_B = 9.0 \ kg/m^3$.
The effect - called cabbaling - will have to be taken into consideration when assuming sea-water as "incompressible", especially in the low temperature range. In the present example we
may conclude that $\partial\rho/\partial x_i \neq 0$, and hence - according to Eq.(4.1.3)
- that $\partial v_i/\partial x_i \neq 0$.

With example 4.1.1 in memory we shall in the following
confine ourselves to the cases where the fluid may be assumed
"incompressible", i.e. where the salinity plays the major role
in determining the density or, where the variations in temperature are within such limits that the coefficient of expansion
(α, see appendix) is quasi-constant. With this limitation, the
local equations of continuity read

$$\frac{\partial v_i}{\partial x_i} = 0 \tag{4.1.4}$$

and

$$\frac{d\Delta}{dt} = \frac{\partial \Delta}{\partial t} + \frac{\partial (\Delta v_i)}{\partial x_i} = 0 \tag{4.1.5}$$

respectively, where $\rho = \rho_R(1+\Delta)$ has been introduced. The depth integrated continuity equations are illustrated by examples in the following.

Example 4.1.2

We consider a fjord (Fig. 4.1.2) with a fresh water runoff R from a river located at the upstream end. On its way out of the fjord, the fresh water is continuously diluted by sea-water from below, for instance due to wind generated entrainment.

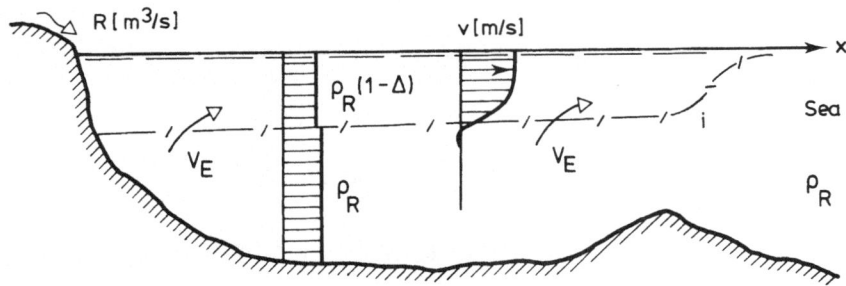

Fig. 4.1.2 A fjord with entrainment (higly distorted scale).

The continuity equation for the total discharge Q states:

$$\frac{dQ}{dx} = V_E B \tag{4.1.6}$$

or in integrated form

$$Q = R + \int_0^X V_E B \, dx \tag{4.1.7}$$

(where B is the interfacial width of the fjord).

Accordingly, we have for the reduced mass flux

$$\frac{d(\Delta Q)}{dx} = 0 \qquad\qquad (4.1.8)$$

or integrated

$$\Delta Q = \Delta_0 R \qquad\qquad (4.1.9)$$

(where Δ_0 is the fresh water mass deficit).

Eqs. (4.1.8) and (4.1.9) state that in a flow with pure entrainment, the reduced mass flux is conserved. If it is the salinity which determines Δ (most common), Eq. (4.1.9) may be rewritten in the salinities to yield (see equation A2)

$$Q(S_{sea} - S) = R(S_{sea} - S_{river} \ (\simeq 0)) \qquad\qquad (4.1.10)$$

An example: The following quantities have been measured: $R = 10^3$ m^3/s, $S_{SEA} = 35\ ^0/_{00}$, S in the upper layer at the mouth = $33\ ^0/_{00}$. Hence, the discharge out of the fjord in the upper layer amounts to

$$Q_{out} = \frac{R \times S_{sea}}{S_{sea} - S_{out}} = 17.5 \times 10^3 \ m^3/s$$

which means that a compensating flow of $(17.5 - 1) \times 10^3$ m^3/s must take place in the lower layer.

The combined continuity equations for volume and mass are often called "The Knudsen relations", in veneration to the Danish oceanographer Martin Knudsen [1900], who applied the relations to get an estimate of the outflow from the Baltic Sea, based on salinity measurements at the Darss-sill and in the Sound.

Another interesting consequence of the constant reduced mass flux is that the critical superelevation $(\Delta y)_c$ at the mouth of the fjord is independent of the entrainment in the fjord, which can be visualized by the following rearrangement

$$\mathbb{F}_{\Delta,c}^2 \simeq 1 = \frac{Q^2}{\Delta g B^2 y_c^3} = \frac{(\Delta Q)^2}{g B^2 (\Delta y)_c^3} \qquad\qquad (4.1.11)$$

and hence

$$(\Delta y)_c = \sqrt[3]{\frac{(\Delta Q)^2}{gB^2}} \qquad\qquad (4.1.12)$$

Exercise 4.1.1

The dense bottom current shown in Fig. 4.1 has a constant rate of entrainment V_E [m/s]. This "uniform" flow is characterized by having a constant densimetric Froude number.

Show that (Δy) and V are independent of the distance x.

Evaluate the depth variation $y(x)$ with the dimensionless entrainment V_E/V as a parameter

Result: $y = y_0 + \dfrac{V_E}{V} x$

Calculate the work to be done against gravity per unit time (the effect) in order to mix the entrained water evenly over the depth of the dense bottom current. The velocity V may be taken constant with depth

Result: $1/2 \; \Delta \rho_R \; g \; y \; V_E \; \cos \phi$

4.2   The equations of motion

The basic equation of motion is the Reynolds' equation, which expresses the balance of forces on a confined volume of mass in a flowing turbulent fluid:·

$$\frac{\partial}{\partial t}\left(\bar{\rho}\,\bar{v}_k\right) + \frac{\partial}{\partial x_j}\left(\rho\left(\bar{v}_k\,\bar{v}_j + \overline{v'_k\,v'_j}\right)\right) + \frac{\partial \bar{p}}{\partial x_k}$$

$$- \bar{\rho}g_k - \bar{\rho}\nu\,\frac{\partial^2\,\bar{v}_k}{\partial x_j\,\partial x_j} = 0 \qquad (4.2.1)$$

(where $^-$ means time average and $'$ turbulent fluctuation).

In integrated form, the Reynolds' equation represents the well-known momentum equation. It often serves the purpose of introducing the reduced pressure $p_\Delta = p - p_R$ (conf. chapter 2). In the reduced pressure system the pressure and gravity terms read:

$$\frac{\partial \overline{p}}{\partial x_k} - \overline{\rho} g_k = \frac{\partial (\overline{p}_\Delta + \overline{p}_R)}{\partial x_k} - \rho_R (1 + \overline{\Delta}) g_k =$$

$$\frac{\partial \overline{p}_\Delta}{\partial x_k} - \overline{\Delta} \rho_R g_k \qquad\qquad (4.2.2)$$

i.e. $\overline{p}$ is replaced by the reduced pressure $\overline{p}_\Delta$ when $\overline{\rho}$ is replaced by the reduced mass $\Delta \rho_R$.

The practical use of the momentum equation does not differ significantly from the way in which we treat open channel flow. This is illustrated by the following examples.

Example 4.2.1

We return to the stationary dense bottom current on a gentle slope I_0, i.e. subcritical flow is presumed, see Fig. 4.2.1.

The depth integrated momentum equation in the flow direction reads

$$\frac{\partial}{\partial x_1} \{ \frac{1}{2} \Delta \rho_R g y^2 \cos \phi + \rho_R \alpha' V^2 y \} - \rho_R V_E u_i$$

$$+ \tau_b + \tau_i - \Delta \rho_R g y I_0 = 0 \qquad\qquad (4.2.3)$$

The first two terms are the socalled "reaction force", i.e. the sum of the pressure and the momentum contributions. Term number three is the momentum caused by the entrained mass.

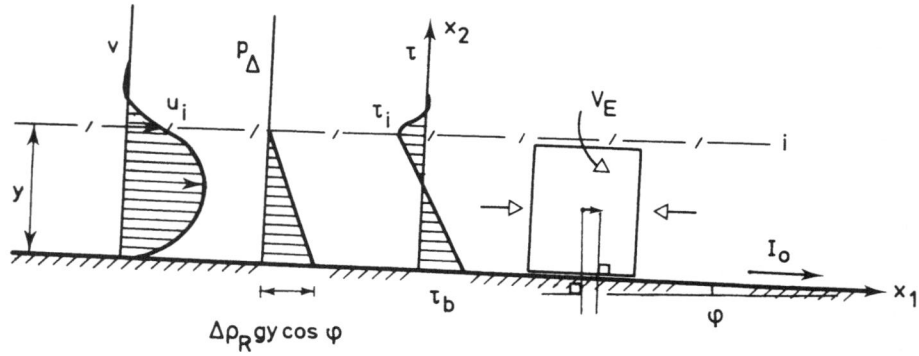

Fig. 4.2.1 *Infinitely wide dense bottom current (without Coriolis effect) with stationary, subcritical flow.*

The next two terms are the shear stresses on the bottom and the interface, respectively, and the last term is the gravitational force. Later - in chapter 8 - it is verified by order of magnitude arguments that equation (4.2.3) degenerates to the similar well-known relation applicable in free surface flow.

$$\tau = \tau_b + \tau_i = \Delta\rho_R \, g \, y \, I_0 \quad (\mathbb{F}_\Delta << \mathbb{F}_{\Delta,c}) \qquad (4.2.4)$$

In supercritical flow (i.e. $\mathbb{F}_\Delta > \mathbb{F}_{\Delta,c}$), equation (4.2.4) does not hold.

Example 4.2.2

The vertical two-dimensional buoyancy jet/plume sketched in Fig. 4.2.2 is created by a line buoyancy flux into an initially stagnant, ambient fluid of higher density.

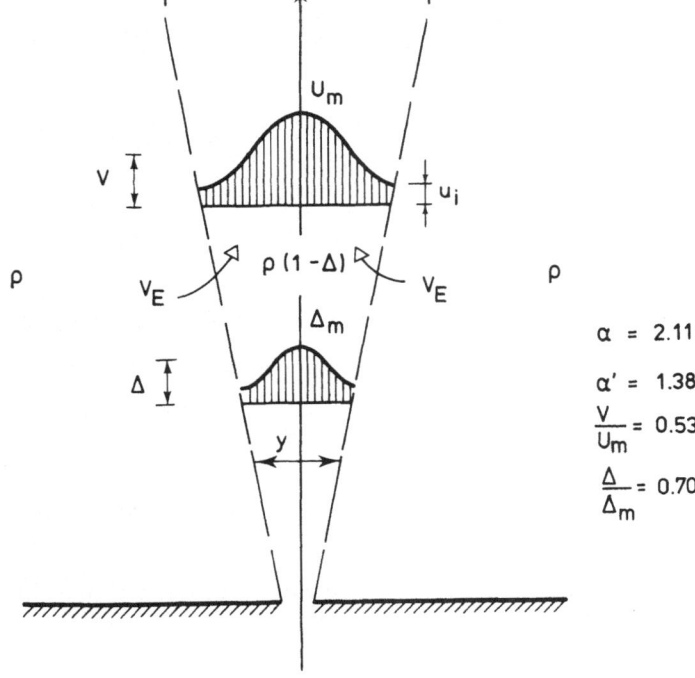

$$\alpha = 2.11$$
$$\alpha' = 1.38$$
$$\frac{V}{U_m} = 0.53$$
$$\frac{\Delta}{\Delta_m} = 0.70$$

Fig. 4.2.2 Rising two-dimensional jet/plume.

The lateral-integrated vertical component of the momentum equation for the jet/plume states per unit width

$$\frac{\partial}{\partial s}(\rho \alpha' \, Vq) - 2\rho \, u_i \, V_E + 2\tau_i = \Delta\rho gy \qquad (4.2.5)$$

where the first term is the momentum of the basic flow, the second term the momentum due to entrainment (from both sides), the third term the interfacial shear stress, and the last term the buoyancy.

If we apply the momentum equation for the ambient fluid, we simply get

$$\tau_i = \rho \, u_i \, V_E \qquad (4.2.6)$$

which means that equation (4.2.5) may be reduced to

$$\frac{\partial}{\partial s}(\rho \alpha' \, Vq) = \Delta\rho gy \qquad (4.2.7)$$

The momentum equation is especially simple for the pure jets (i.e. when $\Delta = 0$)

$$\frac{\partial}{\partial s} \, (Vq) = \frac{\partial}{\partial s} \left(\frac{q^2}{y}\right) = -\frac{q^2}{y^2} \frac{\partial y}{\partial s} + \frac{2q}{y} \frac{\partial q}{\partial s} =$$

$$V^2 \left[-\frac{\partial y}{\partial s} + 4 \frac{V_E}{V} \right] = 0 \qquad (4.2.8)$$

Hence, for a pure jet the spreading is related to the entrainment by

$$\left(\frac{\partial y}{\partial s}\right)_{jets} = 4 \frac{V_E}{V} \simeq 0.36 \qquad (4.2.9)$$

(where the figure 0.36 is experimentally - theoretically determined).

Jets and plumes are discussed further in chapter 12.

_____

An alternative equation of motion is the work energy equation, which is extensively used in ordinary hydraulic calculations. It is outside the scope of the present lecture notes to outline the integrated energy equation for stratified flows, as it is a rather time consuming and tedious affair. Nevertheless, the energy equation - which in fact consists of two equations: one for the mean flow and one for the turbulence - is the key to understanding the physics of stratified miscible flows and will therefore briefly be discussed. A detailed discussion is reported in Bo Pedersen [1980].

The energy equation for the mean flow originates from the Navier-Stokes equation, known from the basic course in hydrodynamics

$$\frac{\partial}{\partial t}\left(\frac{1}{2}\overline{\rho}\ \overline{v_k}\ \overline{v_k}\right) + \frac{\partial}{\partial x_j}\left(\frac{1}{2}\overline{\rho}\ \overline{v_k}\ \overline{v_k}\ \overline{v_j}\right) + \overline{v_k}\ \frac{\partial}{\partial x_j}\left(\overline{\rho}\ \overline{v_k'\ v_j'}\right) +$$

$$\frac{\partial}{\partial x_k}\left(\overline{p}\ \overline{v_k}\right) - \overline{\rho}\ g\ \overline{v_3} - \rho\ \nu\ \overline{v_k}\ \frac{\partial^2 \overline{v_k}}{\partial x_j \partial x_j} = 0 \qquad (4.2.10)$$

and similarly, the energy equation for the turbulence

$$\frac{\partial}{\partial t}\left(\frac{1}{2}\overline{\rho}\ \overline{v_k'\ v_k'}\right) + \frac{\partial}{\partial x_j}\left(\frac{1}{2}\overline{\rho}\ \overline{v_j}\ \overline{v_k'\ v_k'} + \frac{1}{2}\overline{\rho}\ \overline{v_j'\ v_k'\ v_k'}\right) +$$

$$\rho\ \overline{v_j'\ v_k'}\ \frac{\partial \overline{v_k}}{\partial x_j} + \frac{\partial}{\partial x_k}\left(\overline{p'v_k'}\right) - \overline{\rho'v_3'}\ g - \overline{\rho\nu\ v_k'\ \frac{\partial^2 v_k'}{\partial x_j \partial x_j}} = 0$$

$$(4.2.11)$$

Although terrifying to look at, the energy equation for the mean motion simply states that the change in kinetic energy is due to the work done by the internal and external forces. The first two terms are the local and the convective rate of change of kinetic energy, respectively. The third term is the work done per unit time by the Reynolds' stresses $\left(\rho\ \overline{v_k'\ v_j'}\right)$. This term, which for flows in shallow regions reads

$$v_s\ \frac{\partial \tau}{\partial n} \qquad\qquad\qquad (4.2.12)$$

(where s = flow direction, n = perpendicular to s) is tightly connected to the term ($\tau\ \partial v_s/\partial n$) expressing the production of turbulent kinetic energy (term number four in equation (4.2.11) - shallow flows). Pure mathematics yields the connection

45

$$\text{PROD} = \int_{\Omega} \tau \, \frac{\partial v_s}{\partial n} \, d\Omega =$$

$$\int_{\Omega} \frac{\partial}{\partial n} \, (\tau v_s) d\Omega - \int_{\Omega} \left(v_s \, \frac{\partial \tau}{\partial n} \right) d\Omega \qquad (4.2.13)$$

where an integration has been performed over the volume Ω (= the depth times unit area), see example 4.2.4.

The physical interpretation of Eq. (4.2.13) is: The energy transferred from the mean flow (last term) is partly used for production of turbulent kinetic energy (PROD) and partly transferred to the ambient fluid (first term on the right hand side). The term PROD is essential in the evaluation of the entrainment and is - according to Eq. (4.2.13) - tightly connected to the energy equation for the mean motion. In fact, it is possible to outline the following general expression for PROD by combining the energy equation and the momentum equation. (Bo Pedersen [1980])

$$\text{PROD} = \tau_i \, (V - u_i) + \frac{1}{2} \, \rho \, V_{E,i} \, (\sqrt{\alpha} \, V - u_i)^2$$

$$+ \, \tau_w \, (V - u_w) + \frac{1}{2} \, \rho \, V_{E,w} \, (\sqrt{\alpha} \, V - u_w)^2 \qquad (4.2.14)$$

(where i = interface and w = an interface or a wall). In example 4.2.4 is given a "quick and dirty" proof of Eq. (4.2.14).

There are still three terms to mention in the energy equation for the mean motion, namely the work done by the pressure and gravity forces, and finally the last term which is the viscous dissipation, which may be neglected for high Reynolds' numbers.

In the example to follow we wish to demonstrate in a simple case how the combined energy and momentum equation yields an expression for PROD (the production of turbulent kinetic energy).

Example 4.2.3

We consider a homogeneous lake in the inertial phase just after a constant wind stress has been imposed on the surface, Fig. 4.2.3. For convenience we take the velocity as linearly distributed with depth in the present example.

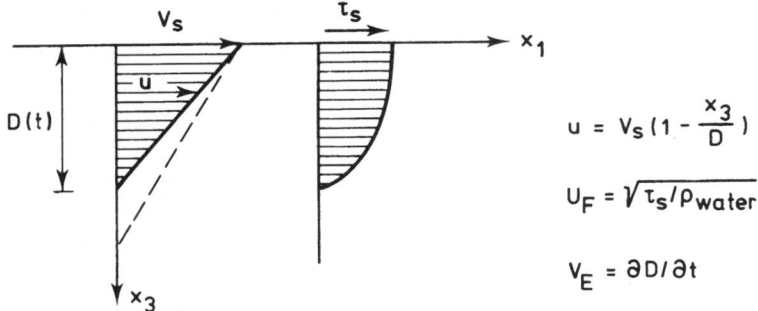

$$u = V_s \left(1 - \frac{x_3}{D}\right)$$

$$U_F = \sqrt{\tau_s / \rho_{water}}$$

$$V_E = \partial D / \partial t$$

*Fig. 4.2.3 Developing boundary layer in homogeneous lake
exposed to a constant wind stress.*

The momentum equation for an infinitesimal element of the
flow reads

$$\rho \, dx_1 \, dx_3 \, \frac{\partial u}{\partial t} = - \, dx_1 \, \frac{\partial \tau}{\partial x_3} \, dx_3 \qquad (4.2.17)$$

where the local rate of acceleration is

$$\frac{\partial u}{\partial t} = V_s \, x_3 \, \frac{1}{D^2} \, \frac{\partial D}{\partial t} = \frac{V_s \, x_3}{D^2} \, V_E \qquad (4.2.18)$$

The distribution of the shear stress is – for a linear velocity profile – determined by Eq. (4.2.17) and Eq. (4.2.18).

$$\frac{\tau}{\rho} = \frac{\tau_s}{\rho} - \frac{1}{2} \, V_E \, V_s \left(\frac{x_3}{D}\right)^2 \qquad (4.2.19)$$

as shown in Fig. 4.2.3. At $x_3 = D$ the shear stress is zero, and
hence Eq. (4.2.19) yields

$$\frac{V_E}{U_F} = 2 \, \frac{U_F}{V_s} \qquad (4.2.20)$$

The energy equation states that the effect per unit area
transferred from the wind ($A_{Ext} = \tau_s \, V_s$) is partly used to increase the kinetic energy of the mean flow ($(1-\eta)A_{Ext}$) and partly used to produce turbulent kinetic energy ($\eta \, A_{Ext}$) which eventually dissipates into heat.

Therefore, the following equation applies

$$\tau_s \, V_s \, (1 - \eta) = \frac{d}{dt} \{\frac{1}{2} \, m < v^2 >\} =$$

$$\frac{1}{6} \, \rho \, V_E \, V_s^2 \qquad\qquad (4.2.21)$$

as the depth integrated velocity squared is

$$< v^2 > = \frac{1}{3} \, V_s^2 \qquad\qquad (4.2.22)$$

for the linear velocity profile. From Eq. (4.2.21) we obtain the following relation

$$1 - \eta = \frac{1}{6} \, \frac{V_E}{U_F} \, \frac{V_s}{U_F} \qquad\qquad (4.2.23)$$

To evaluate PROD we need a value for η which may just be determined by combining the momentum equation (4.2.20) and the energy equation (4.2.23)

$$\eta = 1 - \frac{1}{6} \, 2 \, \frac{U_F}{V_s} \, \frac{V_s}{U_F} = 2/3 \qquad\qquad (4.2.24)$$

and hence

$$\text{PROD} = \frac{2}{3} \, \tau_s \, V_s \qquad\qquad (4.2.25)$$

which states that two-third of the imposed effect is used for production of turbulent kinetic. energy, while only one third is used to increase the kinetic energy of the mean flow.

Exercise 4.2.1

Deduce Eq. (4.2.25) by using the definition of PROD (Eq. 4.2.13).

Example 4.2.4

In the present example we shall make equation (4.2.14) for the PROD probable. We may have a dense bottom current on our mind during the calculations.

The two terms $\tau (V - u)$ are the major contributions to PROD in subcritical flows. In these, τ is linearly distributed - just as in open channel flows, see Fig. 4.2.1. We apply Eq. (4.2.13). The energy transferred per unit area to the neighbouring elements amounts to

$$\int_0^Y \frac{\partial}{\partial n} (\tau v_s) \, dn = - \tau_i u_i - \tau_w u_w = - \tau_i u_i \qquad (4.2.26)$$

(as the velocity at the wall is zero).

The energy transferred from the mean flow to the turbulence is (per unit area)

$$- \int_0^Y v_s \frac{\partial \tau}{\partial n} \, dn = \frac{\tau_w + \tau_i}{y} Vy = (\tau_w + \tau_i) V \qquad (4.2.27)$$

(as $\partial \tau / \partial n = (\tau_w + \tau_i)/y$ is constant). Hence, we find

$$\mathrm{PROD}_{\text{"sub"}} = \tau_i (V - u_i) + \tau_w (V - u_w) \qquad (4.2.28)$$

as postulated above.

The terms $1/2 \, \rho \, (\alpha V^2) \, V_E$ are the major contributions to PROD in supercritical flows. In this flow range the velocity distribution is approximately as illustrated in Fig. 4.2.2 for a jet/plume. For the sake of simplicity we approximate the real velocity distribution with a linearly distributed velocity, see Fig. 4.2.4.

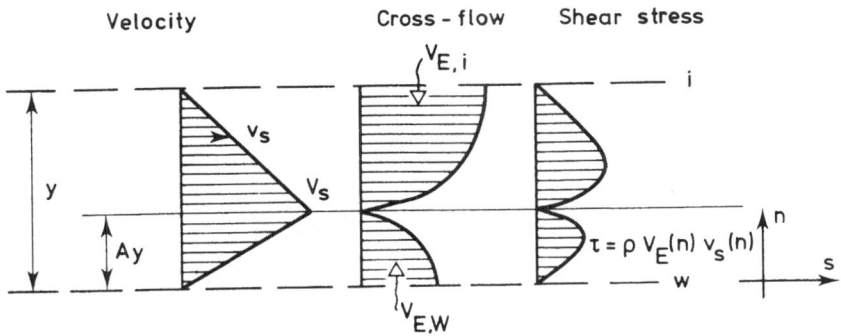

Fig. 4.2.4 *Supercritical flow with an approximated velocity profile. The associated cross-flow and shear stresses are shown as well.*

Assuming similarity in the velocity profiles, we have for reasons of continuity

$$V_{E,i} \; ds \sim V_s (1-A) y$$

$$V_{E,w} \; ds \sim V_s \; A \; y$$

$$\frac{V_{E,i}}{V_{E,w}} = \frac{1-A}{A} \; ; \quad A = \frac{V_{E,w}}{V_{E,i} + V_{E,w}}$$

$$(4.2.29)$$

(where $\sim$ stands for proportional to).

The local cross-flow velocity may be determined by the local continuity equation

$$\frac{\partial v_s}{\partial s} + \frac{\partial v_n}{\partial n} = 0 \qquad (4.2.30)$$

which is integrated to yield (for the lower part of the flow)

$$v_n = V_{E,w} - \int_0^n \left(\frac{\partial v_s}{\partial s} \right) dn = V_{E,w} \left(1 - \left(\frac{n}{Ay} \right)^2 \right) \qquad (4.2.31)$$

i.e. a parabolic distributed cross-flow velocity.

The shear stress distribution can be evaluated by applying the momentum equation. In example 4.2.2 it was shown that the streamwise momentum of a jet is conserved (Eq. 4.2.8), and hence we simply get the momentum from the cross-flow to balance the shear stress

$$\tau = \rho \; v_n \; v_s \qquad (4.2.32)$$

or by introducing Eq. (4.2.31)

$$\tau/\rho = V_{E,w} \; V_s \left(1 - \left(\frac{n}{Ay} \right)^2 \right) \left(\frac{n}{Ay} \right) \qquad (4.2.33)$$

i.e. a third power parabolic shear stress distribution as sketched in Fig. 4.2.4.

Finally, by definition, the production of turbulent kinetic energy is for the present flow situation

$$PROD/\rho = \int_0^Y \left(\frac{\tau}{\rho}\frac{\partial v_s}{\partial n}\right)dn =$$

$$1/4\ V_{E,i}\ V_s\ (1-A)\ \frac{V_s}{(1-A)} + 1/4\ V_{E,w}\ V_s\ A\ \frac{V_s}{A} =$$

$$1/4\ V_s^2\ (V_{E,i} + V_{E,w}) \tag{4.2.34}$$

We may compensate for the crude approximation concerning the velocity distribution by introducing the cross average velocity $V = V_s/2$ and the velocity distribution coefficient $\alpha = \left(\int_0^Y v_s^3\,dn\ /\ V^3 y\right) = 2$. Hence, the result, Eq. (4.2.34), may be expressed in more general terms as

$$PROD_{\text{"super"}} = 1/2\ \rho\ (\alpha\ V^2)\ (V_{E,i} + V_{E,w}) \tag{4.2.35}$$

in accordance with the statement above.

In any flow, being sub- or supercritical, both the contributions outlined are present, only with different importance. Therefore, the total $PROD = PROD_{\text{"sub"}} + PROD_{\text{"super"}}$.

Finally, a discussion of the important energy equation for the turbulence is pertinent. Without any serious loss in the practical use, we may confine ourselves to gradually varying, quasi-stationary, two-dimensional stratified flows (i.e. shallow regions), where the energy equation for the turbulence is reduced to

$$\frac{\partial \bar{e}}{\partial t} + \frac{\partial}{\partial x_j}\ (\bar{v}_j\ \bar{e}) = \tau\frac{\partial \bar{v}_s}{\partial n} - \frac{\partial}{\partial n}\ (\overline{p'\ v_n'} + \overline{v_n'\ e})$$

$$- g\ \overline{\rho'\ v_n'}\ \cos\phi + \rho\varepsilon \tag{4.2.36}$$

where the following new symbols have been used

$$\bar{e} = 1/2\ \rho\ \overline{v_k'\ v_k'} = \text{the turbulent energy per unit volume}$$

$$\varepsilon = \nu\ \overline{v_k'\frac{\partial^2 v_k'}{\partial x_j \partial x_j}} = \text{the dissipation per unit mass.}$$

With reference to Fig. 4.2.5 we perform an integration of the single terms in Eq. (4.2.36). The volume of integration Ω is equal to y times a unit area.

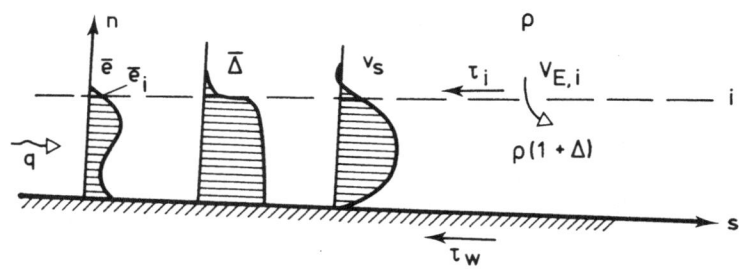

Fig. 4.2.5 Sketch of the distribution of turbulent kine-
            tic energy (ē), reduced density (Δ̄) and velo-
            city (v_s) in a two-dimensional gradually vary-
            ing, stratified flow.

The convective transport of turbulent kinetic energy yields

$$\int_{\Omega} \frac{\partial}{\partial x_j} (\overline{v}_j \, \overline{e}) \, d\Omega = \int_A \overline{v}_j \, \overline{e} \, dA_j =$$

$$\frac{\partial}{\partial s} \left(\delta_e < \overline{e} > q \right) - \overline{e}_i \, v_{E,i} - \overline{e}_w \, v_{E,w} \qquad (4.2.37)$$

where the symbol δ_e stands for

$$\delta_e = \frac{\int_0^y \overline{v}_s \, \overline{e} \, dn}{< \overline{e} > q} \qquad (4.2.38$$

i.e. an energy-flux distribution coefficient which has the order of magnitude one.

(We remember: - time average; < > depth average)

The production of turbulent kinetic energy

$$PROD = \int_0^y \left(\tau \, \frac{\partial \overline{v}_s}{\partial n} \right) dn \qquad (4.2.39)$$

is - as mentioned above - determined by the combined energy and momentum equation, Eq. (4.2.14).

The diffusion out of the flow region

$$\int_0^Y - \frac{\partial}{\partial n} \left(\overline{p' v_n'} + \overline{v_n'\, e} \right) dn \sim 0 \qquad (4.2.40)$$

is neglected as usual.

The buoyancy flux $\overline{g \rho' v_n'}$ is assumed to be associated with the flux of mass due to entrainment, and hence

$$\int_0^Y - g \left(\overline{\rho' v_n'} \right) \cos \phi \, dn =$$

$$- \frac{1}{2} \, \xi_i \, \Delta \rho g y \, V_{E,i} \, \cos \phi - \frac{1}{2} \, \xi_w \, \Delta \rho g y \, V_{E,w} \, \cos \phi \quad (4.2.41)$$

which is recognized to be the rate of change in potential energy per unit time (POT) due to the entrainment, confer with Ch. 2 and exercise 4.1.1. The term $1/2 \, \xi \, y$ is the center of gravity movement in the n-direction of the entrained mass ($\xi = 1$, when Δ is uniformly distributed).

Finally, the rate of energy dissipation is just denoted

$$\int_\Omega \rho \, \varepsilon \, d \, \Omega = - \text{DISS} \qquad (4.2.42)$$

The total integrated energy equation for the turbulence reads

$$\text{PROD} - q \, \frac{\partial}{\partial s} \left(\delta < \overline{e} > \right) = \text{DISS}$$

$$+ V_{E,i} \left[\frac{1}{2} \, \xi_i \, \rho \Delta g y + \{ \delta_e < \overline{e} > - \overline{e}_i \} \right]$$

$$+ V_{E,w} \left[\frac{1}{2} \, \xi_w \, \rho \Delta g y + \{ \delta_e < \overline{e} > - \overline{e}_w \} \right] \qquad (4.2.43)$$

which may be given the following physical interpretation:

The production of turbulent kinetic energy (PROD) minus the convectively transported part of the same (left hand side of equation 4.2.43) - in short: the energy at disposal locally - is primarily used for dissipation (DISS, i.e. into heat) and secondarily used to increase the potential (POT) as well as the turbulent kinetic energy (KIN) of the entrained mass (V_E). This balance is sketched in the diagram, Fig. 4.2.6.

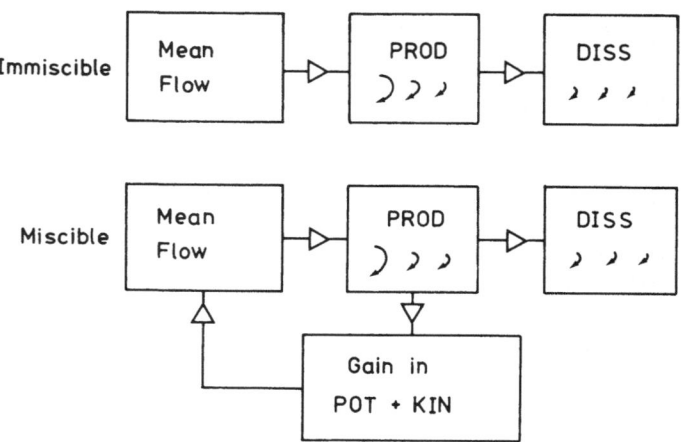

Fig. 4.2.6 Illustration of the energy transport in immiscible and miscible fluids.

The subject for the next chapter is to investigate the efficiency in the energy-transfer-process, i.e. to relate the gain in POT and KIN (the entrainment) to the PROD-term (the mean flow).

5. $\mathbb{R}_f^T$ = THE BULK FLUX RICHARDSON NUMBER

$\mathbb{R}_f^T$ may be defined in two ways:

1) $\mathbb{R}_f^T$ is the ratio between the gain in potential as well as turbulent kinetic energy due to the entrained mass and the energy available for the turbulence, i.e. the production, corrected for the rate of change of the turbulent kinetic energy of the turbulent body.

An alternative definition is,

2) $\mathbb{R}_f^T$ is the efficiency of the recovery (due to entrainment) of the energy produced. Denoting the depth integrated values of the dissipation and the production DISS and PROD respectively gives DISS $\approx$ $(1 - \mathbb{R}_f^T)$ (PROD - Corr), where Corr = the correction stated above.

The basic idea of relating the rate of increase in potential energy to the rate of energy production per unit area, the conventional flux Richardson number $\mathbb{R}_f$, originates from the energy equation for the turbulence (chapter 4). The local value of $\mathbb{R}_f$ is defined as the ratio of the gain in potential energy to the rate of production of turbulent kinetic energy, which for a density current inclined at an angle φ to the horizontal is

$$\mathbb{R}_f = \frac{g \,\overline{\rho' v'_n}\, \cos \varphi}{\tau \dfrac{\partial \overline{v}_s}{\partial n}} \tag{5.1}$$

where hydrostatic pressure distribution perpendicular to the flow direction is assumed. Primarily, we shall be concerned with the depth integrated counterpart of the flux Richardson number,

$$\langle \mathbb{R}_f \rangle = \frac{\displaystyle\int_0^y (g \,\overline{\rho' v'_n}\, \cos \varphi)\, dn}{PROD} =$$

$$\frac{V_{E,i}(\tfrac{1}{2}\xi_i \rho \Delta g y \cos \varphi) + V_{E,w}(\tfrac{1}{2}\xi_w \rho \Delta g y \cos \varphi)}{PROD} \tag{5.2}$$

(see Eq. 4.2.41).

It has often been found that this conventional flux Richardson number is a constant. This is in agreement with an entrainment velocity relative to a representative velocity proportional to the densimetric Froude number squared (i.e. $V_E/V \sim \mathbb{F}_\Delta^2$, see Fig. 6.6).

On the other hand, when the flow is associated with high values of the densimetric Froude number, the entrainment increases at a continuously decreasing rate with the densimetric Froude number, and furthermore the rate of increase in potential energy becomes negligible (just consider a jet where POT is zero). Hence the hypothesis of the constant flux Richardson number has its limit.

In order to eliminate this drawback a new bulk flux Richardson number $\mathbb{R}_f^T$ has been introduced. In a flow situation similar to the one mentioned above (see chapter 4) is

$$\mathbb{R}_f^T = \frac{V_{E,i}\left(\tfrac{1}{2}\xi_i \rho \Delta gy\cos\varphi + \{\delta<\bar{e}> - \bar{e}_i\}\right) + V_{E,w}\left(\tfrac{1}{2}\xi_w \rho \Delta gy\cos\varphi + \{\delta_e<\bar{e}> - \bar{e}_w\}\right)}{PROD - q \frac{\partial}{\partial s}(\delta_e<\bar{e}>)}$$

(5.3)

As the convective term in the denominator normally plays an insignificant role, the denominators for practical use are often identical in the old and in the new flux Richardson numbers. As no detailed knowledge is present on the level of the turbulent kinetic energy, we shall introduce a measure of the degree of turbulence b by the equality

$$\delta<\bar{e}> - \bar{e}_i = \tfrac{1}{2}b\rho(\alpha V^2 - u_i^2)$$

(5.4)

Hence, if we neglect the convective term for a moment, a comparison between the ordinary flux Richardson number $<\mathbb{R}_f>$ and the new $\mathbb{R}_f^T$ is straightforward:

$$\mathbb{R}_f^T \simeq <\mathbb{R}_f> + \frac{\tfrac{1}{2}\rho b\alpha V^2\left(V_{E,i}\left(1 - \frac{u_i^2}{\alpha V^2}\right) + V_{E,w}\left(1 - \frac{u_w^2}{\alpha V^2}\right)\right)}{PROD}$$

(5.5)

As one of the main working hypotheses is that $\mathbb{R}_f^T$ is nearly constant, Eq. (5.5) yields a possibility for establishing the limit for the normal flux Richardson number to be constant. As the normal flux Richardson number is more handy for practical use, it is worthwhile to make a more throughout analysis in order to establish this limit.

The limitation criterion can be depicted from Eq. (5.5) by calculating the ratio of the two numbers:

$$\frac{\mathbb{R}_f^T}{<\mathbb{R}_f>} \simeq 1 + \frac{\frac{1}{2}\rho b \alpha V^2 \left(1 - \frac{u_i^2}{\alpha V^2}\right)}{\frac{1}{2}\xi_i \rho \Delta g y \cos \varphi} \tag{5.6}$$

where, for the sake of clearness, we have restricted ourselves to a case with only one-sided entrainment. Hence, the condition looked for is

$$\mathbb{F}_\Delta^2 = \frac{V^2}{\Delta g y} << \frac{\cos \varphi}{b\left(1 - \frac{u_i^2}{\alpha V^2}\right)} \sim \frac{\cos \varphi}{b} \tag{5.7}$$

which is a simple densimetric Froude number constraint. An alternative expression is obtained, if we take the order of magnitude of bV^2 equal to the friction velocity squared $U_F^2 (= \tau/\rho)$:

$$\mathbb{F}_{\Delta,F}^2 = \frac{U_F^2}{\Delta g y} << \cos \varphi \tag{5.8}$$

Although the knowledge of the parameter b is rather restricted due to lack of measurements, it is possible to make an order of magnitude analysis of the terms in Eq. (5.8). The results are plotted in Fig. 5.1 which only intends to be a comprehensive answer to the practical important question: "In which geophysical phenomena is it allowable to use the constant flux Richardson number concept, and in which has one to use the constant bulk flux Richardson number $\mathbb{R}_f^T$? As illustrated in Fig.5.1 there are a great number of important cases which legitimate the introduction of $\mathbb{R}_f^T$.

From Fig. 5.1 it can be concluded that whether one takes a factor of 10^{-2} (as shown in Fig. 5.1) or 10^{-1} as a measure of much less than (<<), the limit for ordinary flow situations cor-

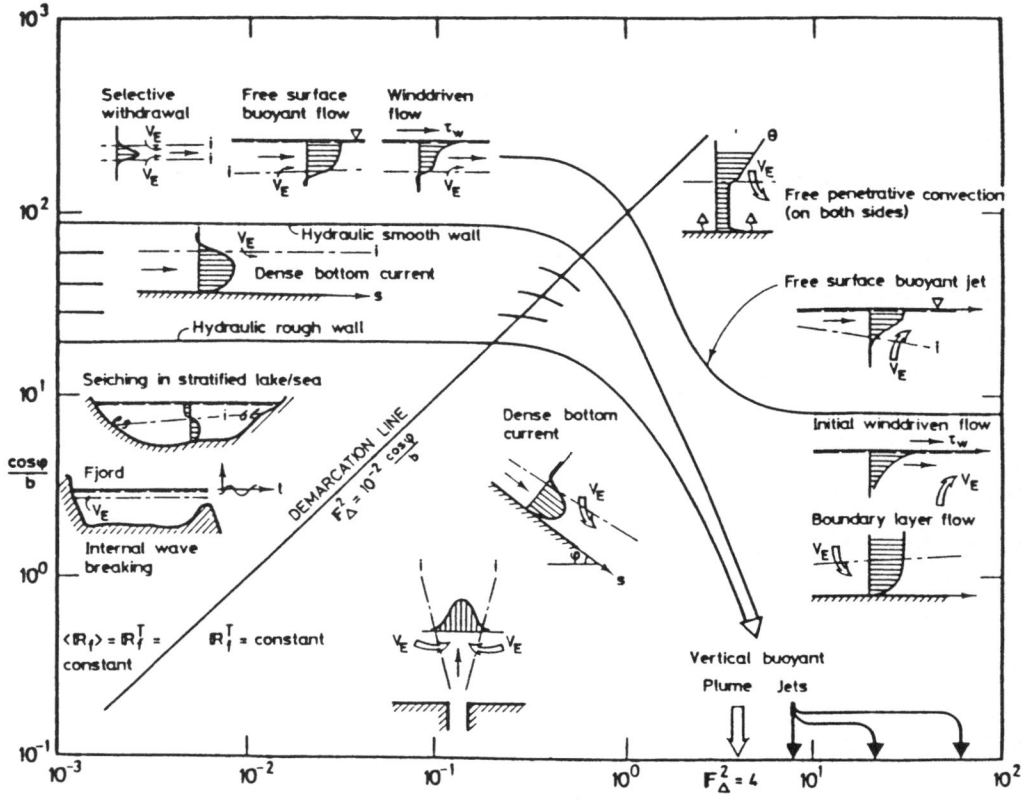

Fig. 5.1 A key to the problem: "When to use $\langle \mathbb{R}_f \rangle =$
$\mathbb{R}_f^T = constant$, and when to use $\mathbb{R}_f^T = constant$?"

$\langle \mathbb{R}_f \rangle = Conventional\ depth\ integrated\ flux$
$\qquad Richardson\ number$

$\mathbb{R}_f^T \quad = The\ bulk\ flux\ Richardson\ number\ intro-$
$\qquad duced\ by\ the\ author$

Symbol ⟋ : cusp-entrainment ⎫
$\qquad$ ⟑ : vortex-entrainment ⎬ see chapter 6
$\qquad\qquad\qquad\qquad\qquad\qquad\qquad$ ⎭

responds to a densimetric Froude number of about 1. Hence the ordinary flux Richardson number can be used meaningfully only when dealing with subcritical flow situations. Furthermore, the limit $\mathbb{F}_\Delta \sim 1$ corresponds to the limit between cusp and vortex generated entrainment, see Ch. 6 for a discussion. This has been indicated by the use of different entrainment symbols in Fig.5.1.

Let us make an order of magnitude analysis of $\mathbb{R}_f^T$. To that end we introduce the "level of turbulence parameter" b, as defined by Eq. (5.4) in Eq. (4.2.43). Hence, definition 2) for $\mathbb{R}_f^T$ yields

$$\frac{DISS}{PROD - q \frac{\partial}{\partial x_1}(b\alpha V^2/2)} = 1 - \mathbb{R}_f^T \tag{5.9}$$

As the dissipation is constrained by the production, it is self-evident that the variation of the bulk flux Richardson number is limited to

$$0 < \mathbb{R}_f^T < 1$$

As $\mathbb{R}_f^T$ degenerates to the normal flux Richardson number for subcritical flow situations, the most commonly mentioned values in the literature can be taken as limits to the variation of $\mathbb{R}_f^T$ in this range:

$$0.04 \leq <\mathbb{R}_f> = \mathbb{R}_f^T \leq 0.1 \text{ for } \mathbb{F}_\Delta^2 << \frac{\cos\varphi}{b} \tag{5.10}$$

For very high values of the densimetric Froude number $\mathbb{R}_f^T$ degenerates into (see equation 4.2.43)

$$\mathbb{R}_f^T = \frac{\frac{1}{2}\rho b\alpha V^2 \left(V_{E,i}\left(1 - \frac{u_i^2}{\alpha V^2}\right) + V_{E,w}\left(1 - \frac{u_w^2}{\alpha V^2}\right)\right)}{PROD - q \frac{\partial}{\partial s}(b\alpha V^2/2)} \tag{5.11}$$

for $\mathbb{F}_\Delta^2 >> \frac{\cos\varphi}{b}$

Unfortunately we have but a few measurements which enable us to evaluate $\mathbb{R}_f^T$ for supercritical flows. If we use the few measurements available, we find

$$0.15 < \mathbb{R}_f^T < 0.2 \quad \text{for } \mathbb{F}_\Delta^2 \to \infty \tag{5.12}$$

Hence, it may be concluded that the bulk flux Richardson number is probably subject to very small variations. The order of magnitude of $\mathbb{R}_f^T$ can be stated as

$$0.04 < 0(\mathbb{R}_f^T) < 0.20 \tag{5.13}$$

for $\quad 0 < \mathbb{F}_\Delta^2 < \infty$

where the high values of $\mathbb{R}_f^T$ apparently correspond to the high values of the densimetric Froude number, i.e. when the entrainment is associated with vortex formation and pairing at the interface, while the low values correspond to the cusp generated entrainment, see Ch. 6.

The problem of the correct variation of $\mathbb{R}_f^T$ cannot be finally solved at the time being due to lack of reliable measurements. Meanwhile, we shall make use of the following values

$$\left.\begin{array}{l} \mathbb{R}_f^T = 0.045 \text{ for } \mathbb{F}_\Delta^2 < \mathbb{F}_{\Delta,cr}^2 \quad (V_{E,cusp}) \\[3mm] \mathbb{R}_f^T = 0.18 \text{ for } \mathbb{F}_\Delta^2 > \mathbb{F}_{\Delta,cr}^2 \quad (V_{E,vortex\ primarily}) \end{array}\right\} \tag{5.14}$$

6. ENTRAINMENT (v_E)

Entrainment can be defined as:

1)   The incorporation of non-turbulent, usually irrotational
        fluid into the turbulent region of the entraining fluid,
or conversely:

2)   The diffusion of the turbulent entraining fluid into the
        non-turbulent ambient fluid.

In geophysical phenomena we are quite often faced with
flow situations where an interface separates two regions of different levels of energy, as for example in estuaries (fjords,
salt water wedges and well mixed estuaries), dense bottom currents in the sea, seiching in a stratified lake, smoke from a
chimney in a wind field, etc. Attempts are constantly being made
to level off this difference in the levels of energy by a mutual
transport of volume from one layer to the other and vice versa,
see Fig. 6.1. Denoting the upward and downward volume fluxes q_u
and q_d respectively, it is common to define the entrainment of
buoyancy (or salt, heat etc.) as the buoyancy flux associated
with the net volume transport from the less turbulent to the
more turbulent region, i.e.

$$\text{buoyancy flux by entrainment} = \Delta(q_u - q_d)g \qquad (6.1)$$

and accordingly to denote the buoyancy flux associated with the
common transport up- and downwards the diffusion of buoyancy,
i.e.

$$\text{buoyancy flux by diffusion} = \Delta q_d \, g. \qquad (6.2)$$

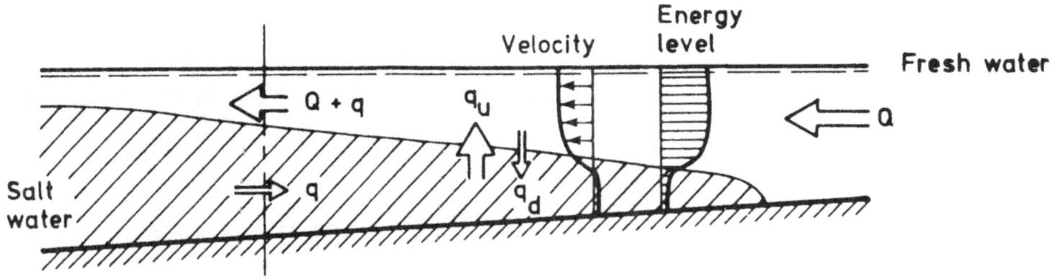

Fig. 6.1   Sketch of the mixing process in a stationary
           salt water wedge.

61

Now, if the lower layer in Fig. 6.1 is at rest and hence
without any turbulence the downward volume flux is nil, and
pure entrainment is present, compare with the definition. An
explanation for this is given later in this chapter.

On the other hand, if the interface is a line of symmetry
(or anti-symmetry) the upward and downward volume fluxes equal
each other, and the term pure diffusion is used for this buoyancy flux mechanism.

Entrainment is - by definition - closely related to turbulence, and consequently it is one of the most outstanding and
complicated problems in geophysics. Although it is an impossible task, we shall try to extract - very briefly - the up-to-
date knowledge of turbulence, but only the part with direct and
obvious relevance to the entrainment process. The reviews of
Roshko [1976] and Laufer [1975] are used extensively. We shall
treat the two-dimensional steady flow only.

Until recently, the most characteristic feature of turbulence was its non-deterministic - its chaotic - eddy motion.

This picture was first radically altered by the Stanford
research group (Kline et al [1967], Kim et al [1971]) who discovered the development in time of a spatially three-dimensional
coherent pattern, called burst. This quasi-ordered structure
moves downstream in the boundary layer. The cycle of events at
any location is intermittent with a broad distribution about a
mean period of $\bar{T} \sim 5\delta/V$, where δ = boundary layer thickness,
V = mean velocity. Much effort has since been spent on research
into these burst-phenomena.

The second epoch-making discovery was made by Brown and
Roshko [1971] who visually observed the existence of the most
beautiful, well-defined large structure rollers or vortices
in the plane, turbulent mixing layer, see Fig. 6.2a.

Once aware of the existence of the coherent large-scale
structure it is relatively easy to visualize it when it is present. On the other hand, it is a rather difficult task to make
informative quantitative measurements of the vortex structure.
This is easily understood if we briefly identify some of the
phenomena to be measured. A detailed description of the vortex
pairing (coalescence) as shown in Fig. 6.2b was first given by

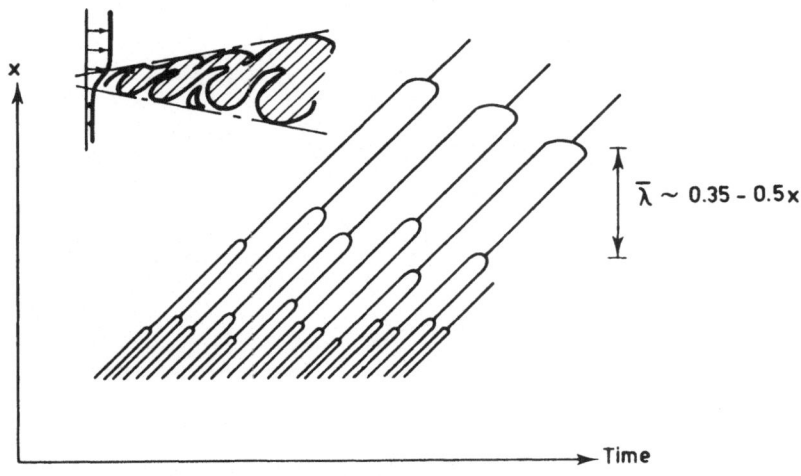

Fig. 6.2a *Large structure rollers in the plane turbulent mixing layer as discovered by Brown and Roshko [1971]. x-t diagram of eddy trajectories in a mixing layer ($\overline{\lambda}$ = average lifespan of an eddy). After Roshko [1976].*

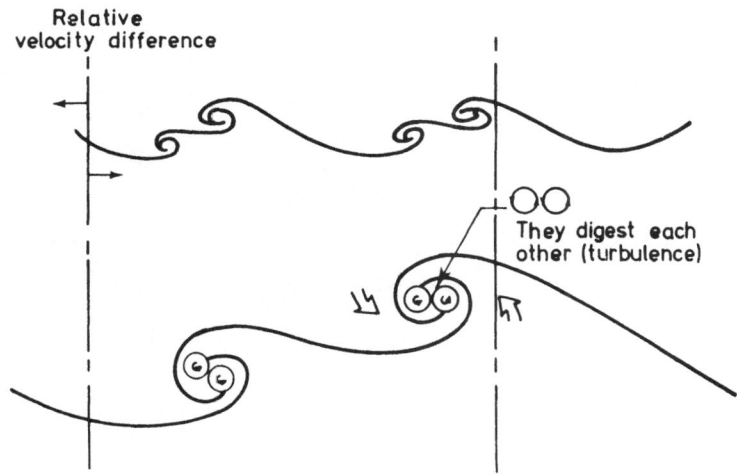

Fig. 6.2b *A moving frame illustration of the successive vortex coalescence process where a pair of vortices are ingested by a single large vortex which, together with its neighbour, is ingested by a single larger vortex which ----.*

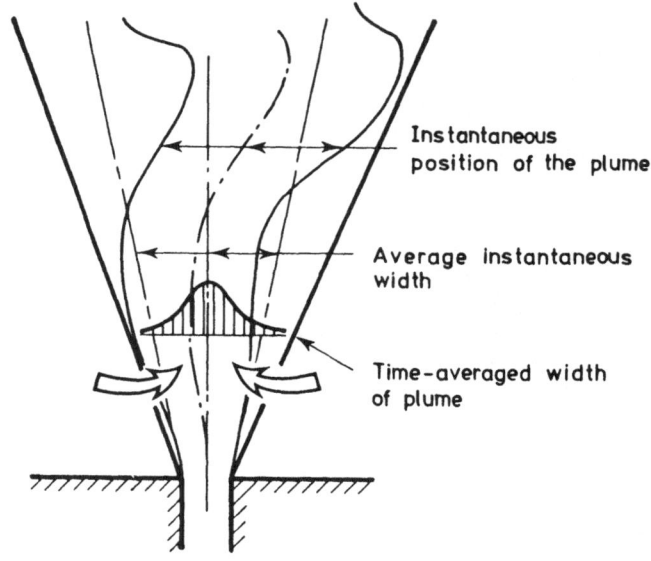

Fig. 6.2c   *Illustration of the meandering effect in a two-dimensional plume (see also Fig. 4 in Part III, Henriksen Haar and Pedersen [1982]).*

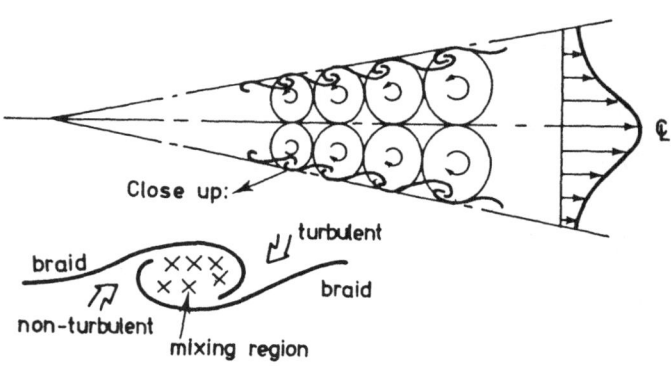

Fig. 6.3   *Schematic vorticity structure in free jet, with close-up of the entrainment process.*

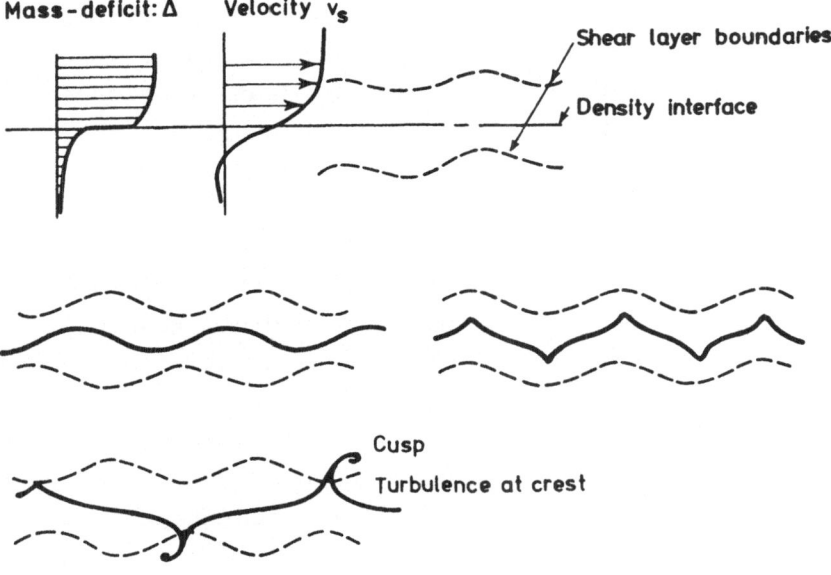

Fig. 6.4   A schematic sequence of a breaking Holmboe
           wave (after Browand and Wang [1972]). Entrain-
           ment by cusp.

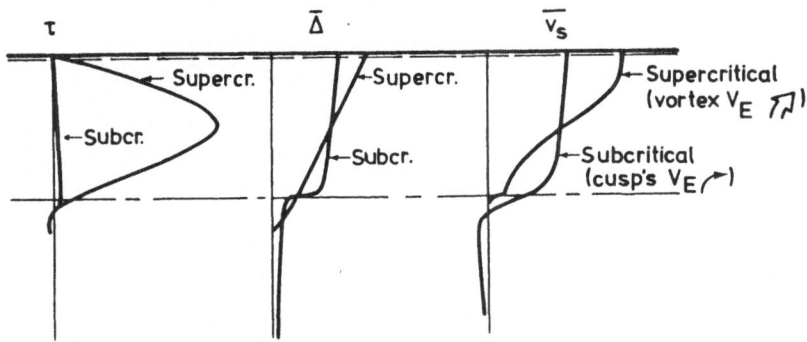

Fig. 6.5   Illustration of the dependence of the dimen-
           sionless density deficit Δ, the Reynolds'
           stresses τ (for equal mean velocities), and
           the velocity v_s respectively on the entrain-
           ment process.

Winant and Browand [1974] (see also Browand and Wang [1972]). As is demonstrated in the x-t diagram and in Fig. 6.2b, the birth of a vortex is associated with the death (coalescence) of two smaller ones. As the two vortices ingested have the same sign of vorticity they digest each other in a turbulence producing process where a mixing of the turbulent/non turbulent fluid takes place too, i.e. entrainment has occurred. This amalgamation process is a statistical process with a standard variation of the eddy lifetime and the eddy spacing, respectively. Furthermore, it is well-known from observations of plumes and free jets that a meandering effect is more the rule than the exception, see part 3 Henriksen, Haar and Pedersen [1982]. Hence, if we try to extract information on the ordered vortex structure from ordinary one- and two-points Eulerian measurements we are faced with the following obstacles (see Fig. 6.2c).

1) Meandering at larger scales which causes intermittency of the signal, 2) statistical variations at scales commensurable with the actual scale which try to flur the phenomenon, and finally, 3) the smaller scale turbulence withing the vortices which contains the integrated effects of the past and hence degrades the ordered signal too. These are some of the reasons why the large-scale ordered structure has not been discovered from the huge amount of turbulence measurements known.

Until now, we have been concerned only with the entrainment process in very high densimetric Froude number flow (insignificant density differences). In the low densimetric Froude number flow range the entrainment becomes very small, but still significant to the flow.

The stability analysis by Browand and Wang [1972] shed light on the reason for this drastical decrease in the rate of entrainment. It showed that the vortex formation and pairing ceased at a critical densimetric Froude number (of the order of magnitude of one). Instead was formed what Browand and Wang denoted Holmboe waves, see Fig. 6.4. The breaking of Holmboe waves is not associated with roll-up and hence the process is less violent. The interface displacement simply grows in amplitude with each succeeding oscillation until breaking at the crest. Breaking occurs on either side of the interface (as was

the case for the coalescing vortices) but due to lack of turbulence in the ambient fluid the entrained fluid is not mixed
further here, but soon after the breaking it will be on its way
back to the interfacial layer, forced by the buoyancy.

It has been suggested (Bo Pedersen [1974a]) to treat diffusion as a two-way entrainment process. If our picture of the
entrainment process as outlined above is basically correct, the
vortex-generated entrainment (Fig. 6.3) as well as the cusp-
generated entrainment (Fig. 6.4) are symmetric around the interface at the moment of breaking. If turbulence is present on one
side only, the entrained fluid is carried away only from the
interface on that side - pure entrainment is present. If, on
the other hand, we have a level of turbulence on both sides, the
double entrained water is removed on both sides, corresponding
to what was denoted entrainment and diffusion above. If the flow
is totally symmetric the mixing process is denoted diffusion.
Hence, with focus on the physical conditions outlined above, it
is more correct to denote the mixing process - and to treat the
transfer of matter as - two-way entrainment.

The change in the physics of the entrainment process about
the critical densimetric Froude number has a remarkable influence on the mean properties of the flow, see Fig. 6.5, where typical subcritical and super-critical flows with identical mean
velocities and mass deficits are illustrated. These radical
changes have a bearing on the flux Richardson number too as discussed in Ch. 5.

After this brief review on the turbulence we face the problem: How do we calculate the entrainment? Our basic hypothesis
is that $\mathbb{R}_f^T$ is a constant for a given flow situation, the constant mainly depending on the densimetric Froude number being
less than or greater than the critical value.

Hence the entrainment in a stationary two-dimensional,
gradually varying density stratified flow can be found by solv-
ing the following equation (see Ch. 5, Eq. (5.3) and (5.4))

$$\mathbb{R}_f^T = \frac{V_{E,i}\left(\tfrac{1}{2}\xi_i \rho \Delta g y \cos \varphi + \tfrac{1}{2}b\rho \left(\alpha V^2 - u_i^2\right)\right)}{PROD - q \frac{\partial}{\partial x_1} (\rho b \alpha V^2 / 2)} \qquad (6.3)$$

As the calculations performed here are to illustrate the use of $\mathbb{R}_f^T$, we have, for the sake of clearness, taken $V_{E,w}$ equal to zero, i.e. subscript w is not an interface.

To proceed from Eq. (6.3) we introduce Eq. (4.2.14), which was a general expression for the PROD valid in the whole range of the densimetric Froude numbers. Hence, Eq. (6.3) can be re-written as

$$\mathbb{R}_f^T = \frac{V_E \frac{1}{2} \xi \rho \Delta g y \left[\cos\varphi + b \, \dfrac{\alpha V^2 - u_i^2}{\xi \Delta g y} \right]}{\tau_i (V - u_i) + \tau_w V + \frac{1}{2} \rho V_E (\sqrt{\alpha} V - u_i)^2 - q \dfrac{\partial}{\partial s} (\rho b \alpha V^2 / 2)}$$

(6.4)

The shear stress terms in Eq. (6.4) are eliminated by introducing the general equations

$$\tau_i = \rho \, \frac{f_i}{2} (U_m - u_i)^2 = \rho \, \frac{f_i}{2} \alpha V^2 \left[\frac{U_m}{\sqrt{\alpha} V} - \frac{u_i}{\sqrt{\alpha} V} \right]^2$$

(6.5)

for the interfacial shear stress (Ch. 7) and

$$\tau_w = \rho \, \frac{f_w}{2} V^2 = \rho \, \frac{f_w}{2} \frac{1}{\alpha} \alpha V^2$$

(6.6)

for the wall shear stress, respectively.

The net flux of turbulent kinetic energy is handled by assuming conservation of momentum (Vq). This approximation is justified by the term being small and of significance only for high values of $\mathbb{F}_\Delta^2$, where momentum is almost conserved. This gives the term

$$- q \frac{\partial}{\partial s} (b \alpha V^2 / 2) = - qV \frac{\partial}{\partial s} (b \alpha V) =$$

$$- \frac{\partial}{\partial s} (b \alpha V^2 q) = - b \alpha V^2 \frac{\partial q}{\partial s} - q \frac{\partial}{\partial s} (b \alpha V^2)$$

(6.7)

or

$$- q \frac{\partial}{\partial s} (b \alpha V^2 / 2) = b \alpha V^2 V_E$$

(6.8)

Hence, Eq. (6.4) can be rewritten as

$$\mathbb{R}_f^T = \frac{\frac{V_E}{V} \frac{1}{2}\left(\frac{\xi \Delta gy}{\alpha V^2}\right)\left[\cos\varphi + \left(\frac{\alpha V^2}{\xi \Delta gy}\right) b\left(1 - \left(\frac{u_i}{\sqrt{\alpha}V}\right)^2\right)\right]}{\frac{f_i}{2}\left[\frac{U_m}{\sqrt{\alpha}V} - \frac{u_i}{\sqrt{\alpha}V}\right]^2\left[1 - \frac{u_i}{V}\right] + \frac{f_w}{2}\frac{1}{\alpha} + \frac{1}{2}\frac{V_E}{V}\left[\left(1 - \frac{u_i}{\sqrt{\alpha}V}\right)^2 + 2b\right]} \quad (6.9)$$

which can be solved with respect to the dimensionless entrainment velocity

$$\frac{V_E}{V} = \frac{2\left[\frac{f_i}{2}\left(\frac{U_m}{\sqrt{\alpha}V} - \frac{u_i}{\sqrt{\alpha}V}\right)^2\left(1 - \frac{u_i}{V}\right) + \frac{1}{\alpha}\frac{f_w}{2}\right]\mathbb{R}_f^T \; \mathbb{F}_\Delta^2}{\cos\varphi + \mathbb{F}_\Delta^2\left[b\left(1 - 2\,\mathbb{R}_f^T - \left(\frac{u_i}{\sqrt{\alpha}V}\right)^2\right) - \mathbb{R}_f^T\left(1 - \frac{u_i}{\sqrt{\alpha}V}\right)^2\right]} \quad (6.10)$$

applicable for the flow situations where only one interface is present. Here the densimetric Froude number squared has been defined as

$$\mathbb{F}_\Delta^2 = \frac{\alpha V^2}{\xi \Delta gy} \quad (6.11)$$

Let us analyse the general behaviour of the entrainment function outlined above. In the subcritical flow range $\cos\varphi \sim 1$ is much larger than the other terms in the denominator. Hence, Eq. (6.10) yields

$$\frac{V_E}{V} \simeq 2\left[\frac{f_i}{2}\left(\frac{U_m}{\sqrt{\alpha}V} - \frac{u_i}{\sqrt{\alpha}V}\right)^2\left(1 - \frac{u_i}{V}\right) + \frac{1}{\alpha}\frac{f_w}{2}\right]\mathbb{R}_f^T \; \mathbb{F}_\Delta^2 \; ; \; \mathbb{F}_\Delta^2 < 1 \quad (6.12)$$

which shows the commonly accepted $\mathbb{F}_\Delta^2$ dependence, as illustrated in Fig. 6.6. To a certain degree the entrainment depends on the wall shear stress and hence on $f_w/2$. The friction coefficient for the wall can be determined by the Colebrook and White formulae (see Engelund and Bo Pedersen [1982]):

$$\sqrt{\frac{2}{f}} = 6.4 - 2.45 \, \ln\left(\frac{k}{R} + \frac{4.7}{\mathbb{R}\sqrt{f}}\right) \quad (6.13)$$

where k = the Nikuradse sand-roughness
     R = the hydraulic radius
    $\mathbb{R}$ = the Reynolds number (based on V, R, and ν).

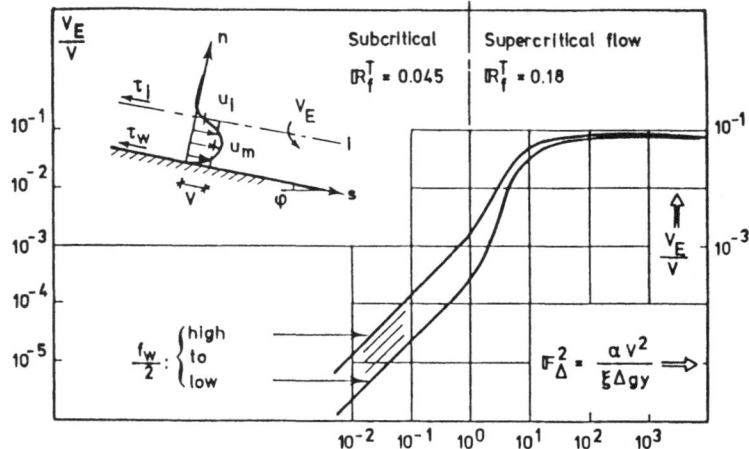

Fig. 6.6 *Illustration of the general shape of the entrainment function (Eq. 6.10) in stationary,
gradually varying density flow.*

For supercritical flow conditions, Eq. (6.10) gives in the
very limit of $\mathrm{I\!F}_\Delta^2 \rightarrow$ infinite (or just $\cos\varphi = 0$) a constant entrainment value with the order of magnitude

$$o\left(\frac{V_E}{V}\right) \sim 0.1 \text{ for } \mathrm{I\!F}_\Delta^2 \rightarrow \infty \tag{6.14}$$

As the interfacial shear stress is now the dominating one, the
curves representing the variation in the wall-shear stress are
brought closer to each other as $\mathrm{I\!F}_\Delta^2$ grows, as shown in Fig. 6.6
where continuous curves have been drawn in the transition re-
gion.

The entrainment function as described by Eq. (6.10) (for
flows with only one interface) is highly dependent on the velocity and the density profiles respectively, so in order to establish an entrainment function which depends only on one parameter,
namely $\mathrm{I\!F}_\Delta^2$, we have to look at some specific examples, i.e. to
classify the flows. This has been done in the following chapters
where a number of simple density currents are treated.

In some cases we have used another approach than the one
outlined above, with due respect to the traditions, but the calculations are all based on the assumption that $\mathrm{I\!R}_f^T$ is a constant,
namely 0.045 for subcritical and 0.18 for supercritical flows
respectively.

7. INTERFACIAL SHEAR STRESS (τ_i)

The shear stress at the boundaries of a flow (including
the interface) is of crucial importance to the macro- as well
as to the microstructure of the flow. The macrostructure - by
which we mean the average velocity, the depth et cetera - is
determined by the momentum equation or the energy equation. The
microstructure - by which we mean the turbulent properties of
the flow - is determined by the conservation equation for the
kinetic energy, the mass flux et cetera. In all these equations,
the shear stress plays a central role by governing the velocity
and depth variations in the flow direction and by determining
the mixing processes including the entrainment.

There is a great number of approaches to this extremely
important problem. The determination of the interfacial shear
stress is normally a delicate affair, because it differs radically from the shear stress generated at fixed boundaries. Hence,
if all the values reported in the literature are plotted uncritically against some more or less relevant parameters, the scatter will be so large that nearly any hypothesis may "not be in-
validated" by the "observed" values.

We do not claim to have found the true facts of the case,
but we feel that the data presented here were selected and treated with so much care that they represent a reliable set of data,
where quality - and not quantity - plays the most important role.

When the flow is *laminar* the local shear stress τ is related to the velocity gradient by Newton's fomula

$$\frac{\tau}{\rho} = \nu \frac{\partial u_s}{\partial n} \tag{7.1}$$

where ν = kinematic viscosity (a material constant).

Knowing the boundary conditions we may, for the laminar
case, evaluate a relation between the interfacial shear stress
and the overall properties of the flow

$$\frac{\tau_i}{\rho} = \frac{f_i}{2} (U_m - u_i)^2 \quad ; \quad \frac{f_i}{2} = \frac{const}{\mathbb{R}e, i} \tag{7.2}$$

$$\mathbb{R}e, i = \frac{(U_m - u_i)(y - y_0)}{\nu} \tag{7.3}$$

where f_i = the interfacial friction factor

U_m-u_i = velocity difference (between maximum and inter-
           facial values)

$y-y_0$ = distance (between the points for U_m and u_i).

The reference velocity and length, respectively, in the
Reynolds number may of course be chosen arbitrarily.

In the case of a *turbulent* flow it is common to relate the
local shear stress (the Reynolds stress) to the velocity gradient by

$$\frac{\tau}{\rho} = K_M \frac{\partial u_s}{\partial n} \tag{7.4}$$

where K_M is the eddy viscosity (not a material constant).

It is common practice to relate the interfacial shear
stress to the velocity squared, as for example done by the relation used here

$$\frac{\tau_i}{\rho} = \frac{f_i}{2} (U_m - u_i)^2 \tag{7.5}$$

Again the choice of reference velocity and length respectively is free, but has, of course, a great bearing on the behaviour of the friction factor $f_i/2$.

The set of formulae outlined above has been used in ordinary open channel flows and in pipe flows with great success for
years (U_m-u_i, $y-y_0$ replaced by V, R = hydraulic radius). From
these flows we have the experience that the friction factor for
turbulent flow is determined by the formulae by Colebrook and
White:

$$\sqrt{\frac{2}{f}} = 6.4 - 2.45 \ln\left(\frac{k}{R} + \frac{4.7}{\mathbb{Re}\sqrt{f}}\right) \tag{7.6}$$

Hence, in ordinary hydraulics, for fixed boundaries, the
friction factor is given by

$$\frac{f}{2} = \frac{f}{2}\left(\frac{k}{R}, \ \mathbb{Re}\right) \tag{7.7}$$

If we try to make a crude analogy to two-layer stratified flow the Nikuradse roughness k - if present - is determined by the microwave field, which is uniquely related to the depth (experimentally verified). Hence, we may expect the relative roughness in a density stratified flow to be a constant

$$\left(\frac{k}{R}\right) = \text{const} \tag{7.8}$$

From the crude analogy above we expect the interfacial shear stress to be dependent on the Reynolds number only. As the velocity profiles for subcritical and supercritical flows respectively are quite different, see Fig. 6.5, it is important to choose the correct reference velocity in order to make the friction factor independent of the densimetric Froude number.

In Fig. 7.1 we have plotted a series of different flow types for which the knowledge of the interfacial shear stress may be of practical importance.

Our main hypothesis is that the interfacial shear stress may be related to the friction factor f_i by

$$\frac{\tau_i}{\rho} = \frac{f_i}{2} (U_m - u_i)^2 \tag{7.5}$$

where U_m = maximum velocity

u_i = interfacial velocity

which means that out reference velocity is the maximum velocity difference within the flow primarily influenced by the interface, i.e. within the region in which the shear stress varies from the interfacial value to zero. This layer thickness is denoted $(y-y_0)$ and the associated Reynolds number is

$$\mathbb{R}e,i = \frac{(U_m - u_i)(y - y_0)}{\nu} \tag{7.3}$$

The only way in which we may obtain the value of the interfacial friction factor is by inspecting the values obtained in the laboratory and in the field. All laboratory and field data plotted in Fig. 7.2 are evaluated in Bo Pedersen [1980].

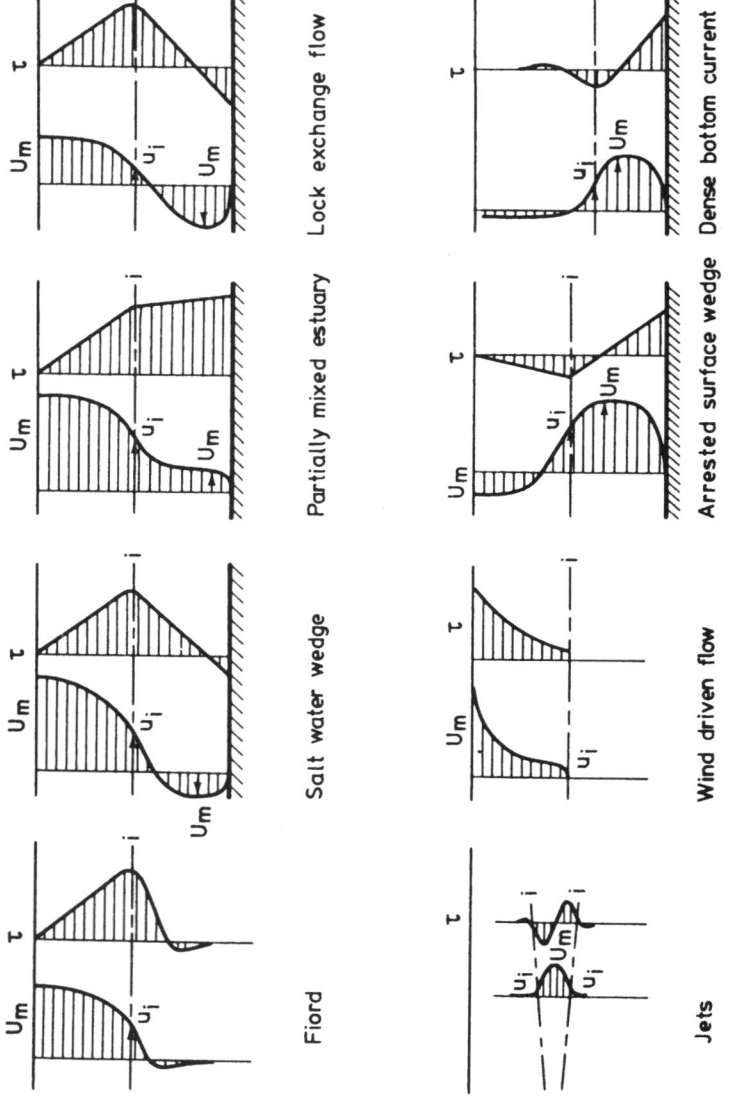

Fig. 7.1 Different types of two-layered stratified flow in which we may
estimate the interfacial shear stress.

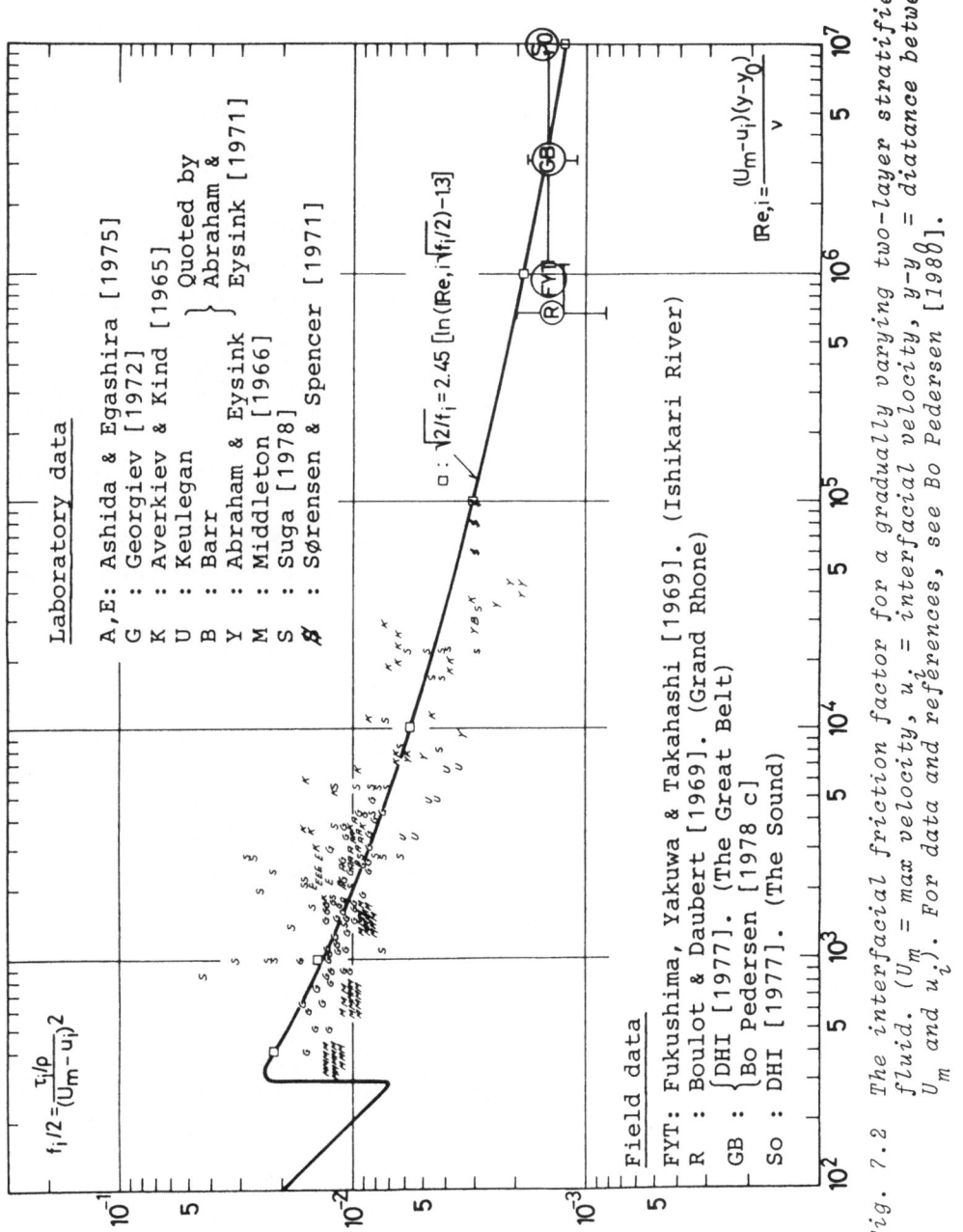

Fig. 7.2 The interfacial friction factor for a gradually varying two-layer stratified fluid. (U_m = max velocity, u_i = interfacial velocity, $y-y_0$ = distance between U_m and u_i. For data and references, see Bo Pedersen [1980]).

The number of field data in our diagram is extremely low compared with the number of data available in the literature, but this is due to the way in which the interfacial shear stress has been evaluated. Most field data stem from overflows which are highly sensible to the influence of the wind.

Hence, if the surface shear stress has not been taken into account, serious errors may arise. Another and much more significant drawback arises from the fact that the flows are treated as quasi-stationary, one-dimensional flows, where due respect to the difference in the friction along the fixed bed and at the interface, respectively, has not been accounted for. Let us illustrate by an example the importance of separating the resistance along the interface and the bed.

Example 7.1

The rivers or straits in which two-layer flow occurs are normally shallow, see Fig. 7.3. The total cross-sectional area for the flow is divided into

$$A = A_i + A_s = b_i y + A_s \tag{7.9}$$

where indices i and s refer to the cross-section where the flow is in contact with the interface and the fixed bottom, respectively. The discharge through sections i and s is divided in such a way that the energy gradients (per unit volume of mass) become equal, i.e.

$$\frac{\tau_i}{\rho g y} = \frac{\tau_s}{\rho g (y/2)} \tag{7.10}$$

where τ is the shear stress and y and (y/2) are the hydraulic radii for the flow through sections i and s, respectively (y/2 corresponds approximately to a triangular cross section). We introduce the friction coefficients

$$\frac{\tau_i}{\rho} = \frac{f_i}{2} (U_m - u_i)^2 \tag{7.5}$$

$$\frac{\tau_s}{\rho} = \frac{f_s}{2} v_s^2 \tag{7.11}$$

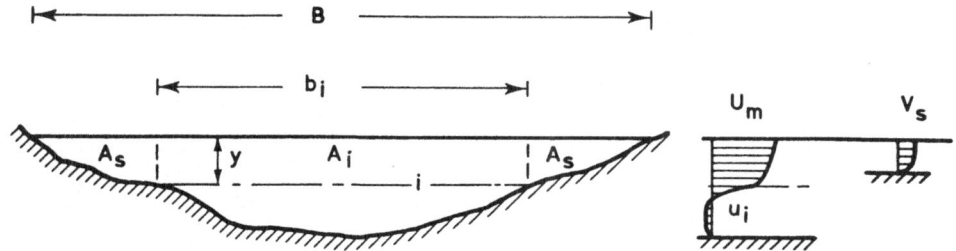

*Fig. 7.3 Cross-section in ordinary shallow river. Index
i refers to the section where the flow is in
contact with the interface, and index s refers
to the section where the flow is in contact with
the fixed bottom.*

in the equation above to yield:

$$\frac{f_i}{2} (U_m - u_i)^2 = 2 \frac{f_s}{2} V_s^2 \tag{7.12}$$

The velocity scale $(U_m - u_i)$ may be related to the mean velocity in the upper layer provided that we have some reliable flow measurements. If we take

$$(U_m - u_i) = c_i V_i \tag{7.13}$$

we obtain the following ratio between the velocities:

$$\frac{V_s}{V_i} = \sqrt{\frac{c_i^2 \ f_i/2}{2 \ f_s/2}} \tag{7.14}$$

Consequently, the cross-sectional mean velocity is

$$V = \frac{Q}{A} = \frac{b_i y}{A} V_i + \frac{V_s}{V_i} \left(V_i - \frac{b_i y}{A} V_i \right) \tag{7.15}$$

or

$$\frac{V}{V_i} = \frac{b_i y}{A} \left(1 - \frac{V_s}{V_i} \right) + \frac{V_s}{V_i} \tag{7.16}$$

This enables us to evaluate the average friction factor for the total upper flow

$$I = \frac{(f/2)V^2}{g(A/B)} = \frac{(f_i/2)c_i^2 V_i^2}{gy} \qquad (7.17)$$

or

$$\frac{f}{2} = \left(\frac{(A/By)}{(V/V_i)^2} c_i^2 \right) \frac{f_i}{2} \qquad (7.18)$$

Fig. 7.4 gives a good illustration of how much the over-all friction coefficient related to the interfacial friction coefficient may vary in a shallow estuary.

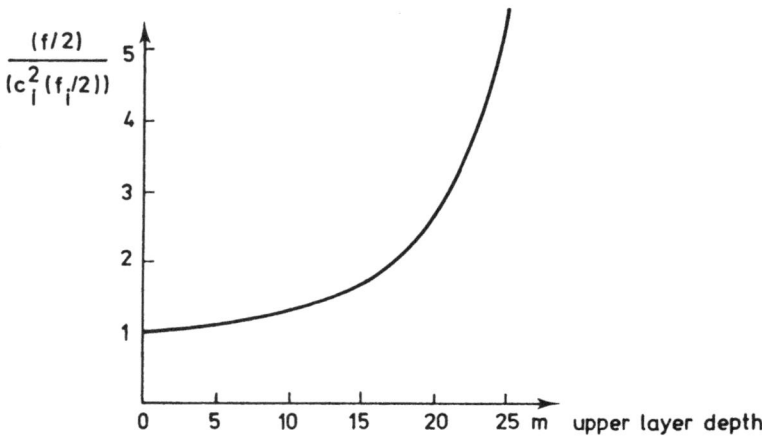

Fig. 7.4 The overall friction factor divided by the interfacial contribution versus the upper layer depth in a shallow estuary (the Great Belt).

The same estuary has been tested by the advanced DHI-mathematical model, see DHI [1977], in which the interfacial friction factor was related to the difference in mean velocity in the upper and the lower layers. If we introduce our findings of this difference in relation to our velocity scale, i.e.

$$\Delta V = 0.61 \, V_{upper} = \frac{0.61}{c_i} (U_m - u_i) = 1.36 \, (U_m - u_i) \qquad (7.19)$$

where the average value of c_i = 0.45 has been introduced, we obtain from the calibration test the estimate

$$\left(\frac{f_i}{2} \right)_{Great \ Belt, \ DHI} = (1.36)^2 10^{-3} \approx 1.8 \times 10^{-3} \qquad (7.20)$$

78

The mathematical model developed by DHI has been used to describe the flow in the Sound, too. This estuary behaves more like a fjord-estuary but with a pronounced velocity in the upper as well as in the lower layers. For this estuary we may estimate on average

$$V_{upper} = 2\ V_{lower} \qquad (7.21)$$

and the constant $c_i = 0.39 = (U_m - u_i)/V_{upper}$. This yields a friction factor for the Sound which is

$$\left(\frac{f_i}{2}\right)_{Sound,\ DHI} = \left(\frac{0.5}{0.39}\right)^2 \times 10^{-3} = 1.6 \times 10^{-3} \qquad (7.22)$$

Consequently, the calibration of the DHI mathematical model which takes the wind, the bed contact, the interface contact et cetera into account gives an interfacial friction factor in good agreement with the other reliable field measurements, see Fig. 7.2.

For practical purposes we may use the following friction factor formula, which is based on the data in Fig. 7.2

$$\sqrt{\frac{2}{f_i}} = 2.45\ \left[\ln\left(\mathbb{R}e,i\ \sqrt{\frac{f_i}{2}}\right) - 1.3\right] \qquad (7.23)$$

$$5 \times 10^2 < \mathbb{R}e,i < 10^7$$

This is shown in Fig. 7.2 to fit the reported data reasonably well. It behaves like an ordinary friction factor for a hydraulic smooth wall, i.e. with an abrupt transition from laminar to turbulent flow and with a decreasing slope as the Reynolds number increases.

8. DENSE BOTTOM CURRENTS

A dense bottom current or a light roof current is the flow created by a source of mass, momentum, and buoyancy flowing into an ambient fluid in such a way that the flow is bounded by the fixed wall and the interface. The dense bottom currents and light roof currents are primarily driven by buoyancy forces.

All oceans, all estuaries, and nearly all lakes and reservoirs which receive water from a river have dense bottom currents. In the ocean and in the estuaries the difference in salinity often plays the most important role in creating the forcing excess density, though temperature and turbidity differences can be present too. In lakes and reservoirs the dense bottom currents are initiated by natural or artificial discharge of a fluid which is heavier than the fluid in the recipient, either due to temperature or to turbidity differences. Consequently there is ample motivation for studying their dynamics. Let us briefly describe some dense bottom currents as they occur in the nature.

Bottom currents in fiords

The hydrodynamics of fiords are a rather complicated matter, see for instance the review paper on fiords by Bo Pedersen [1978b]. Many fiords have one or more sills, i.e. a shallow ridge across the fiord, see Fig. 8.1. This means that the interchange with the adjacent ocean of the water body below the sill level is highly restricted due to blocking effects. A deep water replenishment in a sill fiord can (as an example valid for the west coast of Greenland, for Norway and for Canada) occur in connection with northerly winds, which induce upwelling of denser oceanic water to shallower depths due to the Coriolis force. The current associated is typical dense bottom current starting at the sill level. During its course downwards the density is continuously decreasing with depth. Therefore, at a certain point the densities eventually match each other and the dense bottom current separates from the bottom after having lost its vertical momentum and consequently spreads out horizontally.

This process is - due to the excess of momentum - normally associated with the generation of internal waves, see for instance Hamblin [1977]. The bottom water renewal associated with the

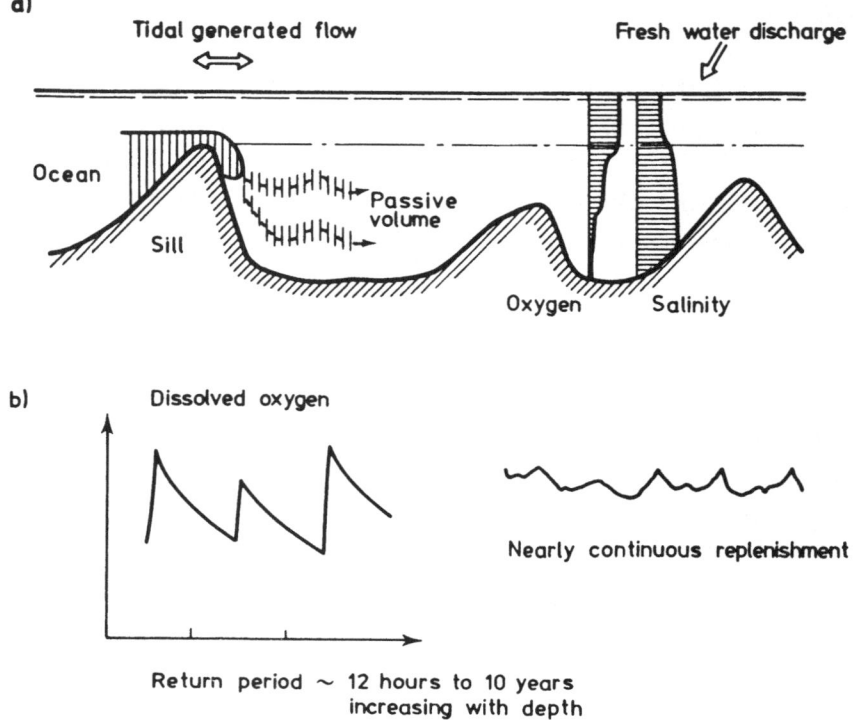

Fig. 8.1 a) A Schematic illustration of a dense bottom
current created by a spill over the sill in a
fiord. b) Examples of oxygen variation in the
bottom water of a sill fiord.

above-mentioned external fiordic forces normally have return periods of, say a year. At the other time limit we have dense water added over the sill with each flood tide. In fiords where
the horizontal extension of the sill is appreciable compared
with the depth, the flow at the sill resembles the arrested
salt water wedge flow. In these cases, the sill density is inversely related to the run-off, and renewals are initiated by
drought, see Edwards and Edelsten [1977], whose measurements
we have included in our entrainment diagram, Fig. 8.10. The sill
of the largest fiord in the world, the Baltic, has a horizontal
extension of a proximately 200×10^3 m in the flow direction
(the Belts and the Cattegat) with a typical depth of 30 meters.
In this fiord the renewal of the deep water is primarily determined by the meteorological conditions over Scandinavia, see

Kullenberg [1977b] and Bo Pedersen [1977]. The deeper bottom water has a return period of 5 to 10 years in connection with a discharge from the Bornholm Basin, increased to about $100 \times 10^3 \, m^3/s$.

Readers interested in further information on the subject "bottom water renewal in fiords" are referred to the literature mentioned above and furthermore to Saelen [1967], Gade [1971, 1973], Stucchi and Farmer [1976], Svendsen [1976] and Møller [1984], where further references can be found.

As the renewal of the bottom water is crucial to the ecological life, it is of great importance to establish the necessary tools for a calculation of this type of currents, which means the entrainment function (chapter 6 and below) and the friction function (chapter 7), respectively.

River entering a lake

As the density of the incoming river water normally differs from the density in the receiving lake, density currents are more the rule than the exception. The density of river water is primarily determined by its temperature and/ or turbidity, where as the density of the ambient fluid, i.e. the lake, is primarily determined by the temperature solely, due to lack of a steady production of turbulence, which is needed for sustaining a significant turbidity. The flow of a river entering a lake displays the same behaviour as the density currents already mentioned in this chapter, except for the boundary condition at the start, see Fig. 8.2. When the river water is denser than the surface lake water, the river proceeds to a position (the plunge line) where it leaves the lake surface and forms a two- or three-dimensional dense bottom current.

In Fig. 8.2 b we have shown a light roof currents as observed by Tesaker [1973] in an ice-covered Norwegian lake.

Katabatic winds in Greenland

As a specific example from the meteorologic field the katabatic winds in Greenland can be mentioned, see Fig. 8.3. The air in direct contact with the cold ground is cooled, which increases

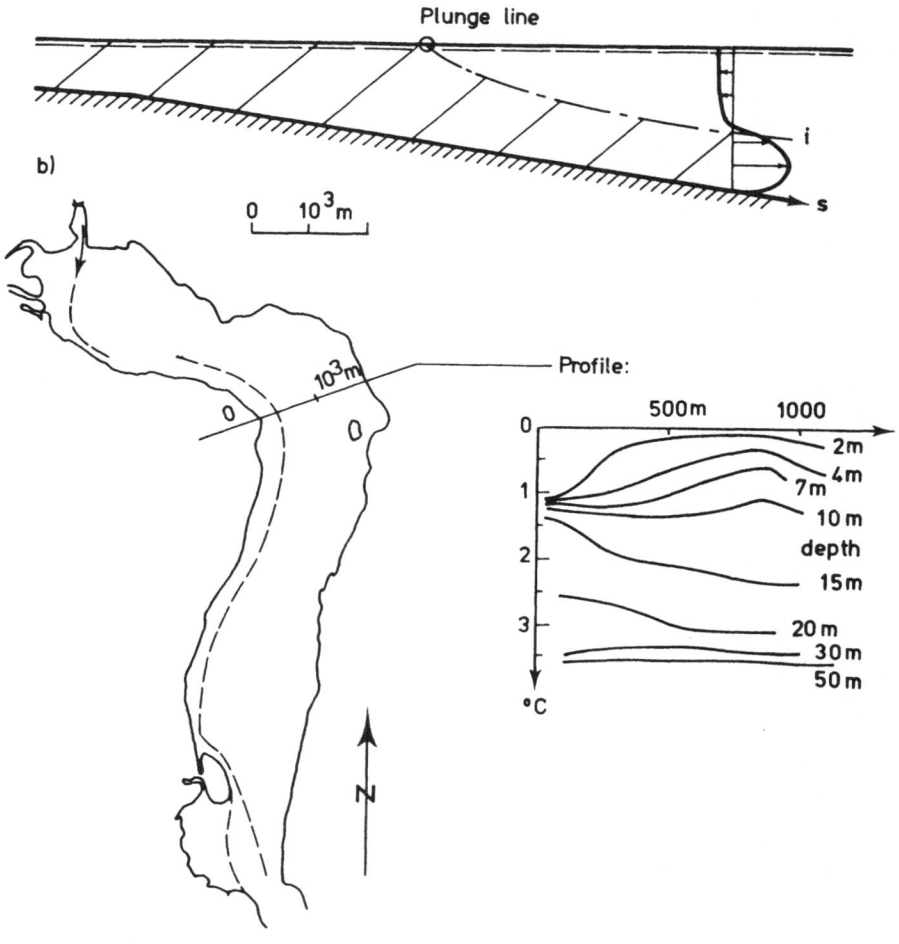

Fig. 8.2   a) Plunge line formed where the dense river
              departs from the surface.
           b) Illustration of light roof current in an
              ice-covered lake on the northern hemis-
              phere.
           From Tesaker [1973].

its density and consequently a non-uniform dense bottom current
is formed.

   In this example we have the combined effects of buoyancy
flux from above (entrainment through the interface) and from below (cooling at the wall) – with which we have confined ourselves
not to deal in the present lecture notes. On the other hand it

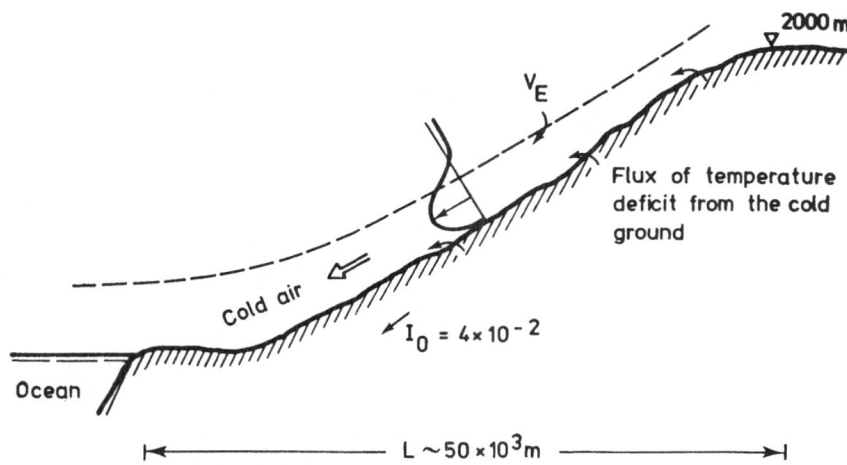

Fig. 8.3 Illustrative example of a katabatic wind at the coast of Greenland. (Not in scala).

is illustrative to look at the order of magnitude of the phenomena (see Fig. 8.3). If we take an ordinary cooling rate of 100 W/m^2 we have a total cooling per unit width of 5 × 10^6 W/m or a reduced mass flux created from the ground

$$\Delta q = \frac{5 \, (MW/m) \times 5 \times 10^{-3} \, (^{\circ}C^{-1})}{10^{-3} \, (MWs/m^{3 \circ}C)} = 25 \; m^2/s \qquad (8.1)$$

If we take this mass flux as representative of the wind, and furthermore assume that the friction factor is, say 10^{-2} we get (see Eq. 4.2.4):

$$I\!F_{\Delta}^{\,2} = \frac{V^3}{\Delta qg} \sim \frac{I_0}{(f/2)} \sim 4 \qquad (8.2)$$

and hence V $\sim$ 10 m/s.

Methane in coal mines

Finally, as an example of a light roof current, the flow of methane accumulated at the roof of a coal mine is illustrated in Fig. 8.4. A detailed knowledge of the flow and the mixing of the methane-air current has an important bearing on safety. It was this aspect of the problem which led Ellison and Turner [1959] to prepare their classical work on turbulent entrainment in stratified flows.

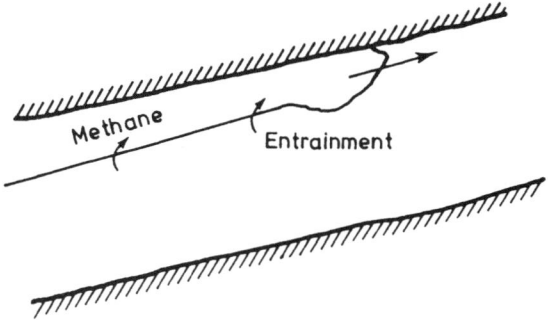

Fig. 8.4 Illustration of light roof current in a coal mine.

In part III further examples on dense bottom currents are given, where the Coriolis effect has been taken into account, see Bo Pedersen [1980b].

Dense bottom currents and light roof currents are, as illustrated above, normally dominated by the buoyancy (gravity) as the driving force and the friction (at the wall and at the interface) as the balancing force. This balance is rapidly reached, which means that the current on a floor with a constant slope very soon reaches a state of equilibrium with a constant densimetric Froude number $\mathbb{F}_\Delta^2$. In the limit of infinite slope, i.e. a vertical wall, the bottom current is transferred to a (half-) falling plume and the roof current to a (half-)rising plume, see Fig. 12.4. Accordingly, dense bottom currents and light roof currents in the equilibrium state are associated with densimetric Froude numbers in the range of

$$\text{nil} < \mathbb{F}_{\Delta,\text{DBC}} \ ; \ \text{LRC} \ < \ \mathbb{F}_{\Delta,\text{plume}} \qquad (8.3)$$

As $\mathbb{F}_{\Delta,\text{plume}}$ is well within the supercritical flow range, we may conclude with reference to chapter 6, that the entrainment process for dense bottom currents and light roof currents on a steeply sloping bottom is associated with vortex engulfing at the interface, while the density currents on a slightly sloping floor has an entrainment caused by interfacial wave breaking. Another aspect of the density currents having this range of densimetric Froude numbers is that in some cases (supercri-

tical flow) the gain in turbulent kinetic energy of the entrained water is appreciable, which implies that the ordinary flux
Richardson number is insufficient as a "closing equation" for
the entrainment process. Therefore, for supercritical dense bottom currents or light roof currents the use of the bulk flux
Richardson number is necessary for deriving a diagnostic equation
for the entrainment, while subcritical dense bottom currents or
light roof currents can be treated with sufficient accuracy by
means of the ordinary flux Richardson number concept (see Fig.
5.1).

In Fig. 8.5 a and b typical velocity and density deficit
distributions, respectively, are shown in a supercritical dense
bottom current as measured by Wilkinson [1970]. In Fig. 8.6 a
and b the similar distributions are shown for a subcritical flow
situation as measured by Georgiev [1972] and is estimated, respectively. Two conclusions can be drawn from the figures:

i)   the general shapes of the profiles are quite different for
       sub- and supercritical flow situations, respectively,
       while, on the other hand,

ii)   the velocity and the density deficit distributions are
       rather insensitive to a variation of the densimetric
       Froude number, provided that we are within the sub- or
       the super-critical regime, respectively.

The general tendency for the interfacial velocity to increase with decreasing densimetric Froude number can be seen in
Fig. 8.5 a, although the scatter in the measured values is large
(this is to be expected due to the intermittent nature of the
interface). For practical uses, the distribution coefficients
for the velocity and the pressure respectively may therefore be
taken as constants in each regime. From the figures we calculate

$$
\left.
\begin{array}{l}
\alpha = 1.84 \quad , \quad \alpha' = 1.28 \\[4pt]
\langle p/\Delta\gamma \rangle \approx 1/2\,(2/3)\,y
\end{array}
\right\} \quad \mathbb{F}_\Delta > \mathbb{F}_{\Delta,cr} \\[14pt]
\left.
\begin{array}{l}
\alpha = 1.12 \quad , \quad \alpha' = 1.04 \\[4pt]
\langle p/\Delta\gamma \rangle \approx 1/2\,y
\end{array}
\right\} \quad \mathbb{F}_\Delta < \mathbb{F}_{\Delta,cr}
\qquad (8.4)
$$

where $\langle p/\Delta\gamma \rangle$ = depth average pressure excess (see Ch. 2).

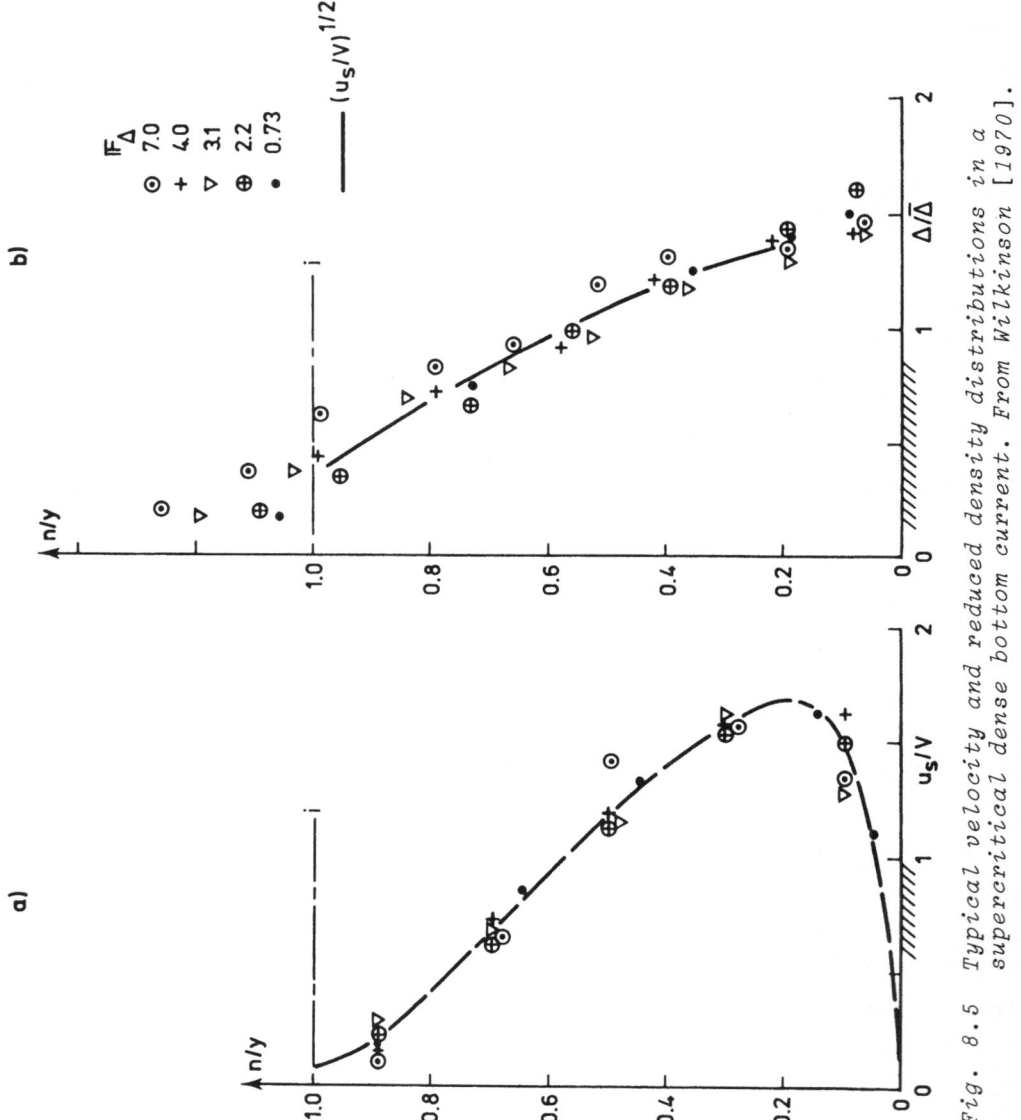

Fig. 8.5 Typical velocity and reduced density distributions in a supercritical dense bottom current. From Wilkinson [1970].

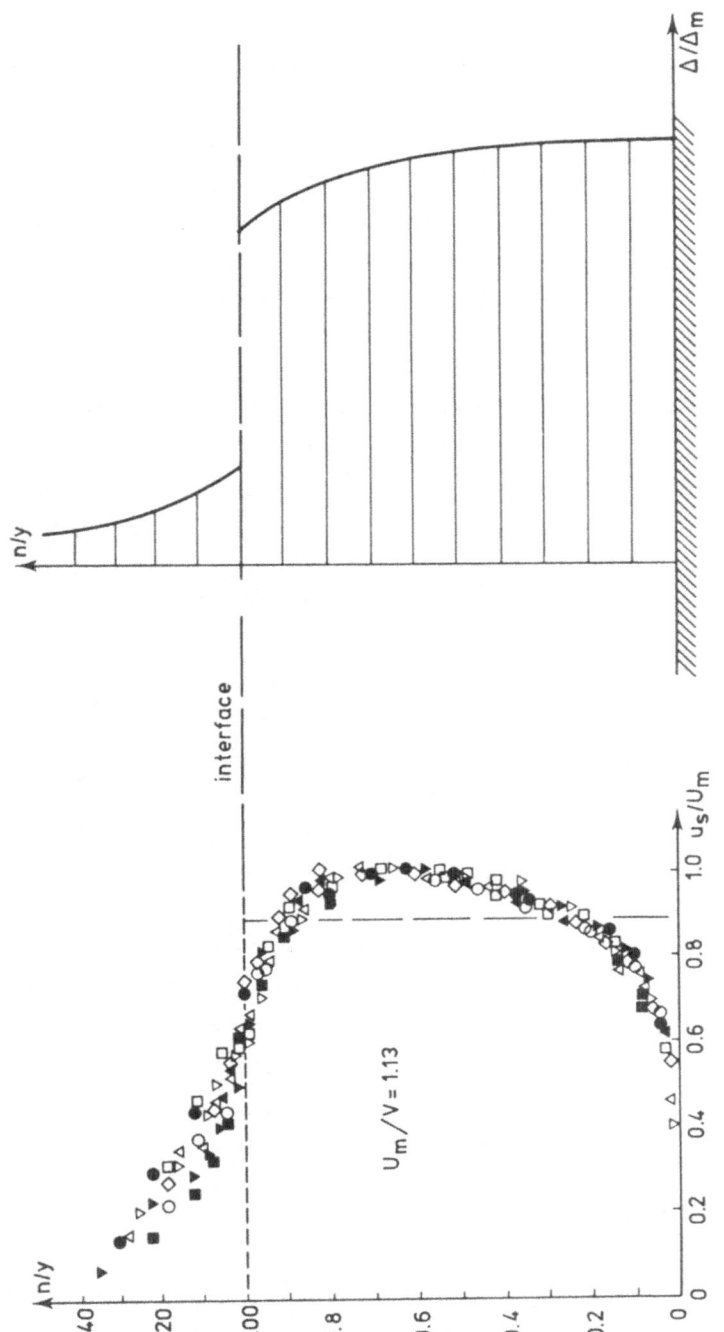

Fig. 8.6 Typical velocity and reduced density distributions in a subcritical dense bottom current. a) from Georgiev [1972]. b) estimated.

The distribution of the shear stress and of the lateral buoyancy flux is of great importance too. Primarily we here have to rely on a qualified guess, as direct measurements are rare. For very high values of the densimetric Froude number we have the intensive jet measurements by Bradbury [1965] as representatives of the shear-stress and the lateral buoyancy-flux distributions, respectively, see Fig. 8.7 a, because the wall shear stress is of the same order of magnitude as the interfacial shear stress, both of which being an order of magnitude less than the maximum shear stress in the flow. Consequently, the existence of a wall has a minor influence in highly supercritical dense bottom currents - in accordance with the statements by Wilkinson [1970]. As the shear stress in the bulk part of the flow is generated by the momentum due to the lateral transport (the entrainment) of mass, we may expect the shear-stress distribution as a function of the densimetric Froude number squared to be as sketched in Fig. 8.8 (see also Fig. 4.2.4). The linear distribution indicated for subcritical flow situations corresponds to a negligible rate of entrainment and a uniform density. Direct measurements of the shear stress distribution in subcritical density currents have been performed by Georgiev [1972], whose measurements confirm the distribution shown in Fig. 8.8. A further conclusion can be drawn on subcritical flows, namely, that the nearly uniform density distribution implies that the buoyancy flux decreases linearly from a maximum value at the interface to zero at the wall. This is radically different from the buoyancy flux distribution in highly supercritical flow, where a nearly parabolic distribution is present as shown in Fig. 8.7.

The general behaviour of a two-dimensional dense bottom current can be described by the momentum equation, which was outlined in example 4.2.1. The result (Eq. 4.2.3) is rewritten here

$$\frac{\partial}{\partial x_1} \left(\frac{1}{2} \Delta \rho g y^2 \cos\varphi + \rho \alpha' V^2 y \right) - \rho V_E u_i$$

$$+ \tau_b + \tau_i - \Delta \rho \, g y I_0 = 0 \qquad (8.5)$$

where subscript R in ρ has been omitted.

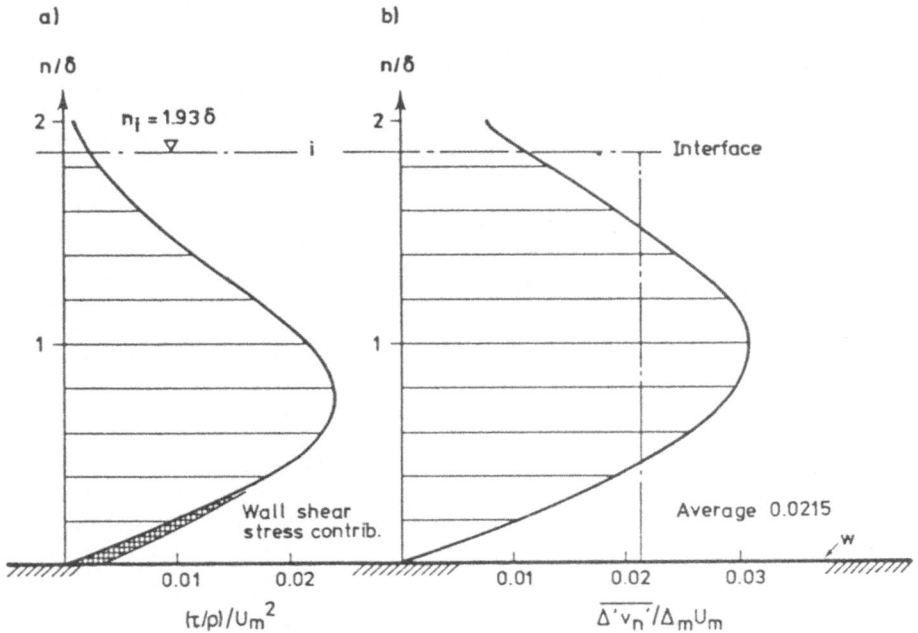

Fig. 8.7 a, b *Typical shear stress and lateral buoyancy flux distributions in a supercritical dense bottom current with negligible wall friction and $\mathbb{F}_\Delta \to \infty$. After Bradbury [1965].*

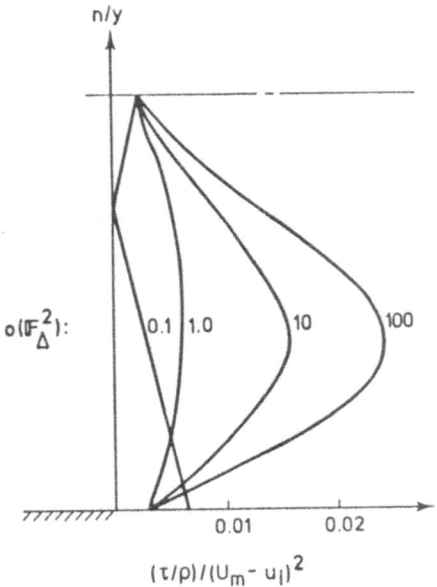

Fig. 8.8 *Rough estimate of the variation of the shear stress in dense bottom currents as a function of the densimetric Froude number squared (high Reynolds numbers).*

To proceed further we recall exercise 4.1.1 from which we learned that (Δy) and V are constant, and that the variation in the depth was

$$\frac{dy}{dx_1} = \frac{V_E}{V} \tag{8.6}$$

Hence, if we further introduce

$$\tau = \tau_b + \tau_i = \rho\, f/2\, V^2 \tag{8.7}$$

in Eq. (8.5) we obtain

$$\left\{\frac{\cos\varphi}{2} + \left(\alpha' - \frac{u_i}{V}\right) \mathbb{F}_\Delta^2\right\} \frac{V_E}{V} + \frac{f}{2} \mathbb{F}_\Delta^2 - I_0 = 0 \tag{8.8}$$

where the first term on the left hand side is an order of magnitude smaller than the other terms involved ($\sim \frac{1}{2} \times V_E/V \sim 0.04\, I_0$ - see later), and therefore may be cancelled. Eq. (8.8) then reads

$$\frac{\Delta g y V}{V^3} = \frac{1}{\mathbb{F}_\Delta^2} = \mathbb{R}i = \frac{f/2 + V_E/V\left(\alpha' - \frac{u_i}{V}\right)}{I_0} \tag{8.9}$$

where $\mathbb{R}i$ is a bulk Richardson number.

In Fig. 8.9 a number of measurements are shown as reported in the literature on Richardson numbers versus the bottom slope. The figure clearly indicates that the data have a trend as indicated by Eq. (8.9). By introducing an upper and a lower limit to the friction factor it is possible to limit the area within which the measurements ought to be located. By definition

$$\frac{f}{2} = \frac{f_i}{2} + \frac{f_w}{2} \tag{8.10}$$

or

$$\frac{f}{2} = \frac{\tau_i/\rho}{(U_m - u_i)^2} \left(\frac{U_m}{V} - \frac{u_i}{V}\right)^2 + \frac{f_w}{2} \left(\frac{U_m}{V}\right)^2 \tag{8.11}$$

where $f_i/2$ has been related to $(U_m - u_i)^2$, in accordance with the

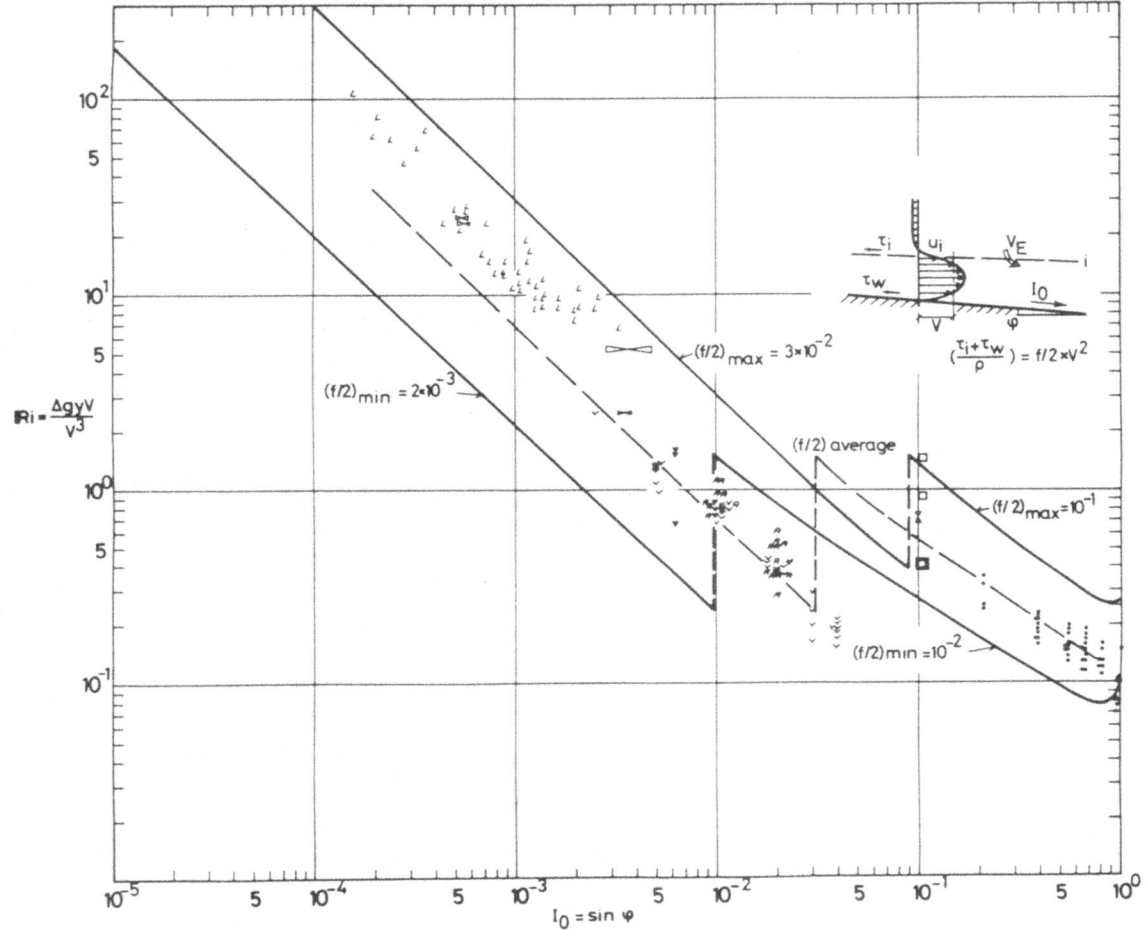

Fig. 8.9 *Laboratory and field data compared with the extreme limits to the bottom slope I_0 versus Richardson number $\mathbb{R}i$. For data and references see Bo Pedersen [1980].*

The points are based on data referred by:

V Middleton [1966] (Laboratory experiments)

· Ellison and Turner [1959] (Laboratory experiments)

□ Edwards and Edelsten [1977] (Deep water renewal of Loch Etive fiord)

Ⅹ Georgeson [1942], average of 107 experiments (mining engineering)

A Ashida and Egashira [1975] (laboratory turbidity currents)

⋈ Smith [1975] and Worthington [1970] (Dense bottom current in the Denmark Strait, example attached, part three).

L Löfquist [1960] (Laboratory measurements), (The depth y is here R = hydraulic radius)

▼ Wilkinson [1970] (Laboratory measurements) y is here = R = hydraulic radius

findings in Ch. 7, and $f_w/2$ has been related to the maximum velocity, compare Fig. 8.5. Consequently, we can estimate $f/2$ as

$$\frac{f}{2} = \frac{\tau_i/\rho}{(U_m - u_i)^2} \frac{1}{3} + \frac{f_w}{2} \qquad \text{for } I\!F_\Delta < I\!F_{\Delta,cr} \qquad (8.12)$$

$$\frac{f}{2} = \frac{\tau_i/\rho}{(U_m - u_i)^2} 2.6 + 2.9 \frac{f_w}{2} \qquad \text{for } I\!F_\Delta > I\!F_{\Delta,cr} \qquad (8.13)$$

which can be combined with the following estimates for the wall and the interface contributions respectively:

$$10^{-3} < \frac{f_w}{2} < 2.5 \times 10^{-2} \quad \text{(see Engelund and Bo Pedersen [1982]}$$
$$(8.14)$$

$$2.5 \times 10^{-3} < \frac{\tau_i/\rho}{(U_m - u_i)^2} < 2 \times 10^{-2} \quad \text{(see Ch. 7)} \qquad (8.15)$$

to yield the possible ranges for $f/2$ as

$$\left.\begin{array}{ll} 2 \times 10^{-3} < \dfrac{f}{2} < 3 \times 10^{-2} & \text{for } I\!F_\Delta < I\!F_{\Delta,cr} \\[3mm] 10^{-2} < \dfrac{f}{2} < 10^{-1} & \text{for } I\!F_\Delta > I\!F_{\Delta,cr} \end{array}\right\} \qquad (8.16)$$

The upper and lower limits to the bulk Richardson number in Eq. (8.9) can then be evaluated by combining Eq. (8.16) with the entrainment function outlined below and shown in Fig. 8.10. (The displacements in the upper and the lower limiting curves are due to the rapid change in the velocity profiles when we pass the critical Richardson number). All available data are nicely located at central positions within the extreme limits in Fig. 8.9, indicating that the data are all meaningful. The only significant deviation stems from the data for a vertical wall, where the dense bottom current is transferred to a falling plume with a radical change in the velocity profile from nearly linear to nearly Gaussian. Therefore, for $I_0 = 1$ one may expect a departure from the theory, see Ch. 12.

In order to get the rate of entrainment we make use of the constant $I\!R_f^T$ concept. This shall be demonstrated for subcritical flow only ($I\!F_\Delta < I\!F_{\Delta,cr}$). In this flow range we have

$$\mathbb{R}_f^T = 0.045 = \frac{POT}{PROD} = \frac{\tfrac{1}{2}\Delta\rho gyV_E}{\tau_i(V-u_i)+\tau_b V} \tag{8.17}$$

(for POT see Ch. 2; for PROD see Eq. (4.2.14))

or

$$\frac{V_E}{V} = \frac{0.09}{\Delta\rho gy}\tau \left[1 - \frac{u_i/V \quad \tau_i/\rho}{\tau_i/\rho+\tau_w/\rho}\right] \simeq 0.072 \, I_0 \tag{8.18}$$

where Eqs. (8.7) and (8.9) and Fig. 8.6 have been used.

A similar - but more complicated - function may be obtained for the super-critical flow range. The results are plotted in Fig. 8.10 and all the measured entrainment values are consistent with the constant bulk flux Richardson number concept, namely $\mathbb{R}_f^T = 0.045$ for $\mathbb{F}_\Delta < \mathbb{F}_{\Delta,cr}$ and $\mathbb{R}_f^T = 0.18$ for $\mathbb{F}_\Delta > \mathbb{F}_{\Delta,cr}$. The data systematically fit the theoretical entrainment function, with a surprisingly small scatter when one considers the diffi- culties in obtaining entrainment data. Even the data of Löfquist [1960], which were performed in a small laboratory flume ($\mathbb{R}e \sim 10^3$ to 10^4) are nearly identical with the data from the Denmark Strait Current (width $\sim 300 \times 10^3$ m, depth ~ 200 m, $\mathbb{R}e \sim 10^6 - 10^7$) calculated in part III, Bo Pedersen [1980b]. The jump be- tween the super- and the subcritical range is due to the change in the bulk flux Richardson number as well as to the change in the velocity profiles.

By a mere coincidence, Eq. (8.18) gives a fairly good estimate of the entrainment of the supercritical flows, too. Therefore, for engineering purposes we may conclude that with- in the scatter corresponding to the up-to-date measuring tech- niques, the entrainment function, Eq. (8.18) may be used for all two-dimensional dense bottom currents.

This has the very great advantage that density currents influenced by the Coriolis force (which are the most common cur- rents in nature) may be treated by means of a dimensionless length scale ℓ defined by

$$\ell = \int_0^s \left(\frac{I_0(s)}{y(s)}\right) ds \tag{8.19}$$

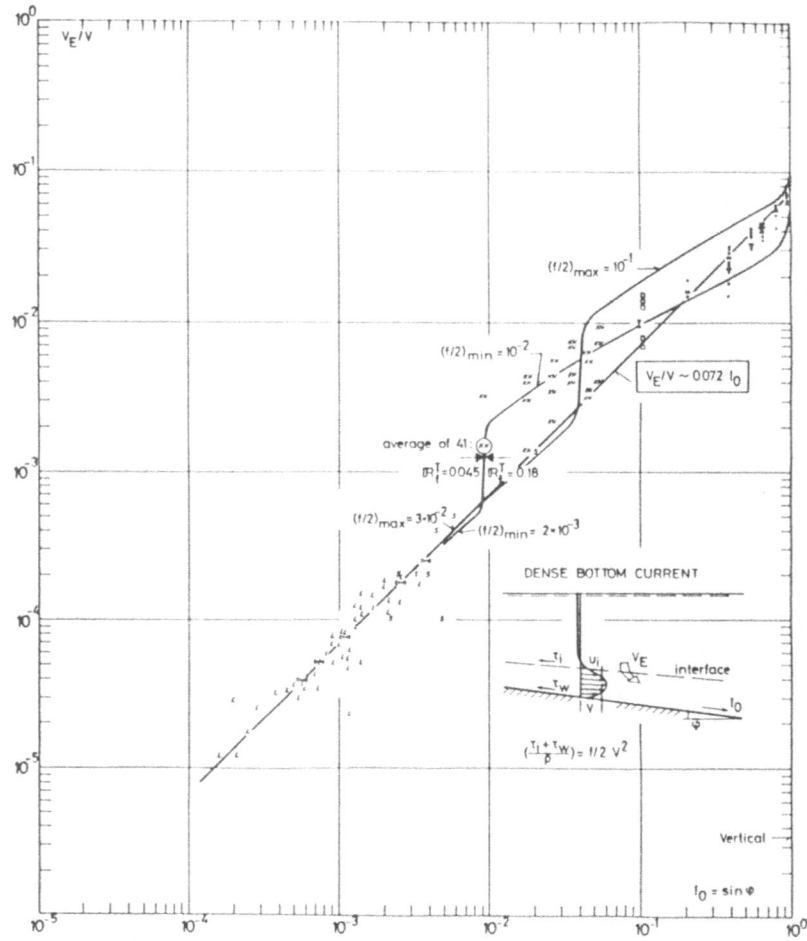

Fig. 8.10   Laboratory and field data on entrainment compared with
            the theory. The lowest and the highest entrainment
            rates are drawn, based on an estimate of the natural
            range of variation of the friction coefficient (f/2).
            For data and references see Bo Pedersen [1980].

The points are based on data referred by:

KH   Kersey and Hsü [1976] (Laboratory   L   Löfquist [1969] (Laboratory expe-
     experiments)   riments)

·   Ellison and Turner [1959] (Labora-   ⋈   Smith [1975] and Worthington [1970]
     tory experiments)   (Dense bottom current in the Den-
                                                 mark Strait, example attached,
□   Edwards and Edelsten [1977] (Deep   Part III)
     water renewal of Loch Etive Fiord)

⌶   Georgeson [1942] (Mining engineer-   S   Suga [1978] (Personal communica-
     ing)   tion. Laboratory experiment)

Using ℓ we overcome the hurdle of the bottom current following a path-line (s) with variable bed slope. This is demonstrated in the above-mentioned example on the Denmark Strait overflow, and elucidated in example 8.1 below.

Example 8.1

We consider a two-dimensional dense bottom current along the path line s. The rate of increase in the discharge per unit width amounts to (by definition)

$$\frac{dq}{ds} = V_E = 0.072 \ I_0 \ V = 0.072 \ \frac{I_0}{y} \ q \qquad (8.20)$$

or by introducing the length scale (Eq. 8.19)

$$\frac{1}{q} \frac{dq}{d\ell} = 0.072 \qquad (8.21)$$

with the solution

$$Q = Q_0 \ \exp \left\{ 0.072 \int_0^s \left(\frac{I_0}{y} \right) ds \right\} \qquad (8.22)$$

or the similar expression for the reduced density ($\Delta Q = \Delta_0 Q_0$)

$$\Delta = \Delta_0 \ \exp \left\{ - \ 0.072 \int_0^s \left(\frac{I_0}{y} \right) ds \right\} \qquad (8.23)$$

The integral $\int_0^s \left(\frac{I_0}{y} \right) ds$ is for a stepwise constant depth simply the drop in elevation ($\int_0^s I_0 \ ds$) non-dimensionalized by the depth y.

By example, a 3 m thick dense bottom current starts in elevation -16 m and ends in elevation -46 m. Hence, its discharge has increased to

$$\frac{Q}{Q_0} = \exp \left\{ 0.072 \ \frac{-16 \ - \ (-46)}{3} \right\} = 2.05 \qquad (8.24)$$

i.e. approximately doubled, and accordingly the mass deficit Δ has been halved.

9. FREE PENETRATIVE CONVECTION

Free penetrative convection is the disorganized movement without a mean velocity created by a source of buoyancy flux flowing into an ambient fluid.

Free convection is often denoted penetrative convection, because the process associated is normally a penetration of a turbulent fluid element into the ambient non-turbulent fluid of stable or neutral stratification. As the buoyancy source in geophysical (natural) free convection stems from the heating or cooling of water/air, a diurnal or seasonal frequency is often present, but as the advancing velocity of the interface is small compared with the velocity fluctuations in the well mixed layer quasi-stationarity is ensured.

Let us illustrate the free penetrative convection in nature by taking three specific examples from different geophysical fields.

The atmospheric inversion rise

The atmosphere quite often exhibits a stable stratification at night. At sunrise, therefore, the sensible heat flux from the ground creates an ustable layer, which is a source of a free penetrative convection. During the daylight hours the well-mixed boundary layer grows by turbulent entrainment to several hundred meters (up to, say two kilometers). The boundary layer is often subject to a horizontal wind too, which makes the computational problem more complicated, and which is the main source of errors when interpreting field data.

The theories and the field measurements of convective boundary layer growth are numerous. In the first theories the downward heat flux at the inversion base was neglected (Ratio = 0), then it was put equal to the surface heat flux (Ratio = 1). This is probably the reason why the entrainment in free penetrative convection has been given indirectly, namely by a certain ratio of the inversion base heat flux to the surface heat flux. The order of magnitude of the Ratio is 0.2. In the present lecture notes the same procedure is used, as it is convenient in weather prediction models and furthermore makes a comparison

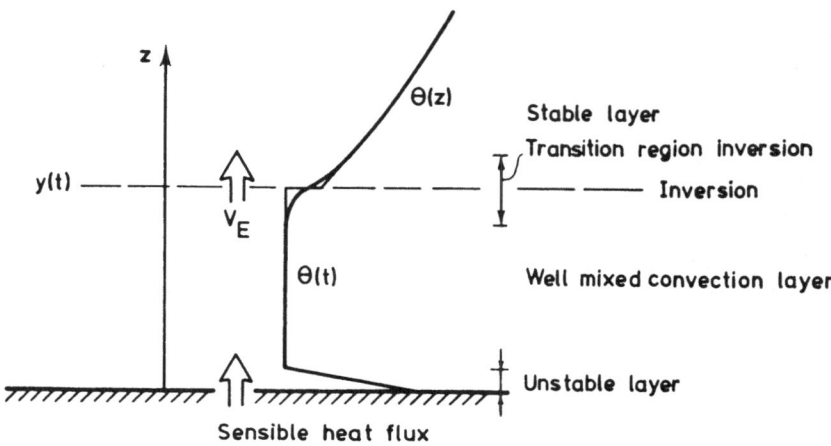

Fig. 9.1   *Sketch of the atmospheric inversion rise*
(θ = potential temperature).

with the data available easy. In some cases, namely where the temperature jump is small or even non-existent (for zero lapse rate above) the constant ratio concept becomes meaningless. These special cases are automatically taken care of by using the constant bulk flux Richardson number, see Ch. 5. We have therefore evaluated an ordinary entrainment function as well, applicable in the whole range of free penetrative convection.

Buoyancy driven circulation

In Ch. 2 (see Fig. 2.7) we discussed the effects of releasing a buoyancy flux at the surface of a stratified estuary (evaporation). Similar phenomena may be encountered in the Arctic in connection with a growing sea-ice sheet ($S_{Sea\ ice} \leq 10$ $^o/oo$, $S_{Sea\ water} \sim 35$ $^o/oo$), which creates a buoyancy flux

$$B_{ice} = V_{ice}\ \beta\ (S_{Sea\ water} - S_{ice})g = \Delta\ V_{ice}\ g \qquad (9.1)$$

where V_{ice} is the ice-growth-rate [m/s].

In the atmospheric inversion rise example above, as well as in the numerical example 2.4, it was assumed that the buoyancy flux created a well mixed convection layer, which had no horizontal density gradient. In nature this assumption is very seldom fulfilled due to the uneven boundary conditions, either

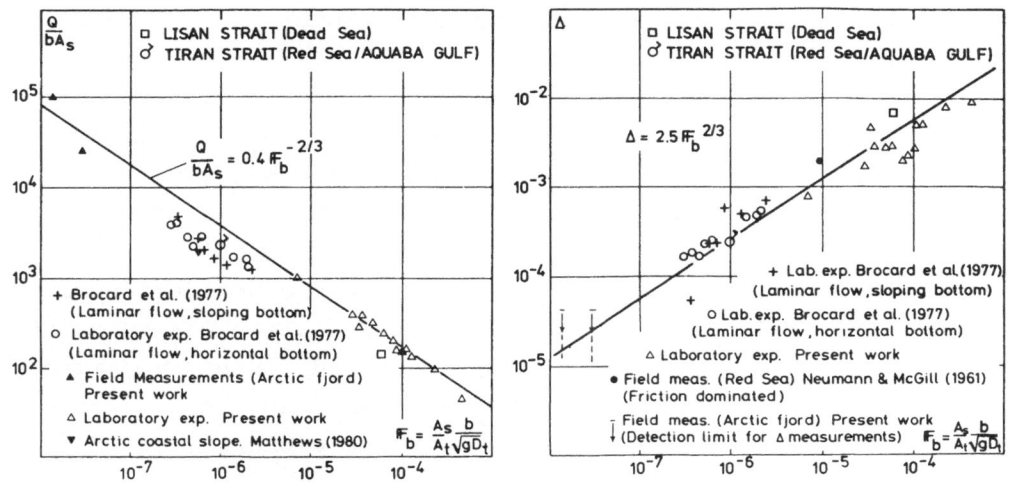

Fig. 9.2 a   Buoyancy driven circulation in a fjord arm.
From Møller [1984]. A_s: surface area of fjord;
A_t: cross sectional area at the sill; D_t:
mean sill depth; b: B/g = the flux of dimensionless mass [m/s].

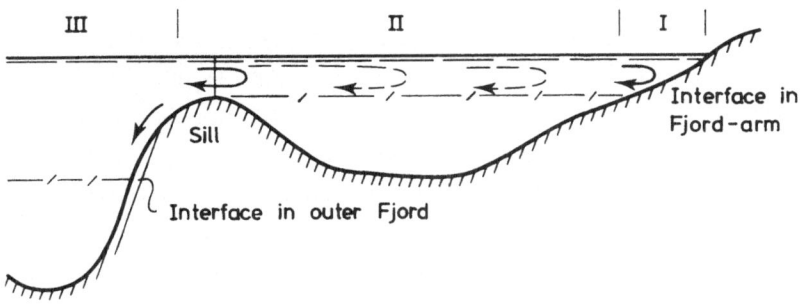

Fig. 9.2 b   Buoyancy driven circulation in an ice-
covered fjord.

the geometric or the physic. Take for instance the fjord arm
sketched in Fig. 9.2 b, subject to an evenly distributed cooling
rate. The associated buoyancy flux creates a density growth rate
inversely proportional to the local depth of the upper layer.
Hence in zone I, which may illustrate the coastal zone, a density driven circulation takes place, which try to compensate for
the density difference. Similarly, the unbalance in pressure

due to the density difference between the inner and the outer fjord creates a "lock-exchange" (example 3.3.3) or "inverse estuarine" circulation. In deep contrast to the fact that this type of flow is one of the basic flow phenomenae encountered all over in the oceans and in the atmosphere, the literature on the subject is extremely sparse. The approach to the problem is normally based on similarity assumptions and dimensional arguments, where no accounts have been taken to the nonstationarity or the energy-balance in the system. One exception can be found in Møller [1984], who has discussed the steady buoyancy driven circulation from an energy point of view. Further he has verified his theory by field and laboratory measurements, see Fig. 9.2 a.

The amplification effect (Q circulation/Q input = b × Area) is tremendous in large estuaries. In the Mediterranean - by example - an evaporation rate of 1.3 m/year creates a circulation current of 1 Sverdrup = 10^6 m^3/s.

The Baltic

The Baltic is the largest fiord in the world with a geometry of length ~ 10^6 m, width 10^5 m, and depth 10^2 m and several sills with depth 10 m < h < 10^2 m. Although it is one of the most intensively investigated oceanographic fields in the world, its very complicated hydrodynamics have left a great number of questions unsolved. One such question is that of the relative importance of the different geophysical processes responsible for the salt and oxygen transfer across the interface. Very briefly, if we look at the central part of the Baltic, see Fig. 9.3 a, the wind is responsible for the production of turbulence in the surface and the deeper water (up- and downwelling and seiching), whereas the cooling of the surface during the autumn and the winter creates free penetrative convection turbulence.

Let us very briefly look at the free convection in the Baltic. The density of the 8 ^{o}/oo saline water is at its maximum when the temperature is about 2°C, which means that the cooling of the surface water in the autumn and the winter creates an unstable surface layer which plunges downwards until it reaches the very stable interface, see Fig. 9.3 c. Local freezing of the

a)

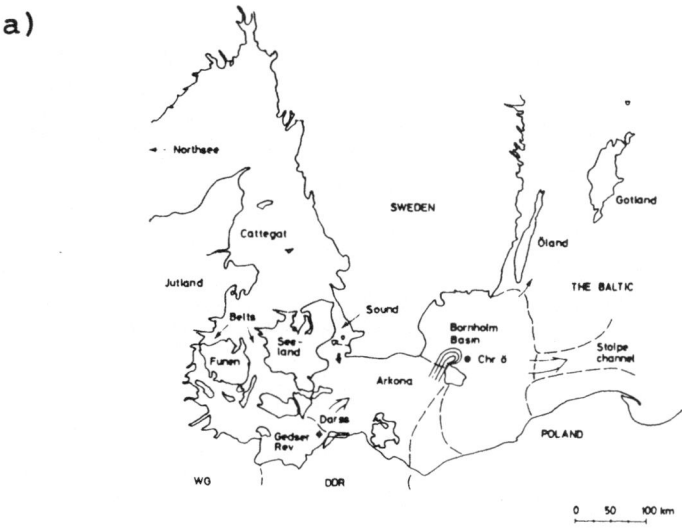

b)

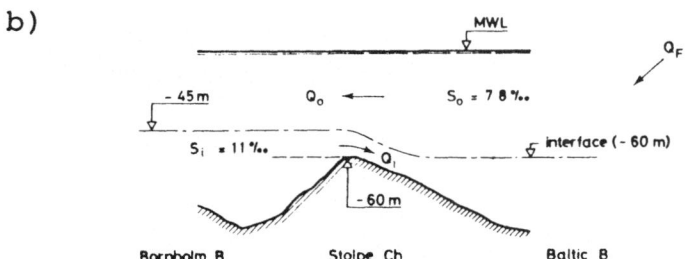

c)

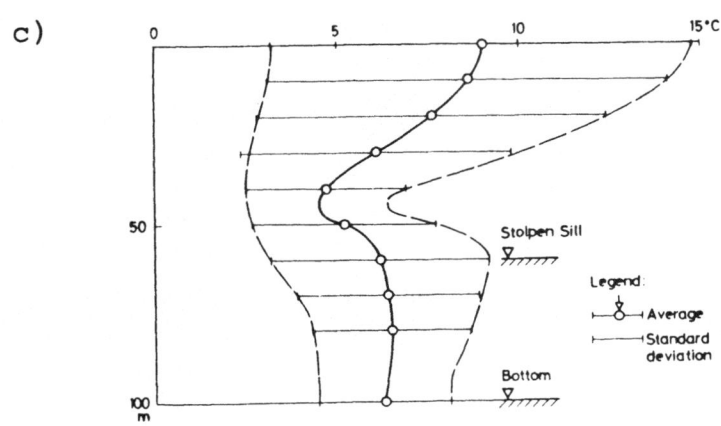

Fig. 9.3   a) Chart of the southern part of the Baltic.
           b) The average location of the interface.
           c) Temperature distribution in the Bornholm
              Basin.
           The figures are from Bo Pedersen [1977c].

water in the southern part of the Baltic is consequently delayed until the total volume of the upper layer of the Baltic has been cooled down to 2°C. This volume of the upper layer amounts to 1.3×10^{13} m^3, and as the surface area is 3.6×10^{11} m^2 the average depth is 36 m. Thus the necessary cooling can be estimated to be (the heat capacity of water being 4.2×10^6 W s/m^3 °C).

$$w = \Delta T \times 36 \times 10^6 \times 4.2 = 1.5 \times 10^8 \, \Delta T \; Wsec/m^2 \qquad (9.2)$$

which for a yearly temperature variation of, say 10°C yields an order of magnitude of the effect

$$o(E_{cooling}) \sim o(E_{heating}) \sim 0.1 \; kW/m^2 \qquad (9.3)$$

This figure gives a total heat budget for the upper layer in the central part of the Baltic, which corresponds to the effect generated by, say 15,000 nuclear power stations. Accordingly, the surface heat flux is of the order of magnitude of

$$o\left((\overline{v_3' \theta'})_w \right) \sim \frac{0.1 \, [kW/m^2]}{4.2 \times 10^3 \, [kWs/m^3\,°C]} \sim 2.4 \times 10^{-5} \left[\frac{m}{s} \, °C \right] \qquad (9.4)$$

Alternatively, this figure can be found directly by the equation of conservation of heat, which simply is

$$(\overline{v_3' \theta'})_w \simeq h \frac{d\theta}{dt} \sim 36 \frac{10}{1.6 \times 10^7} \sim 2.3 \times 10^{-5} \left[\frac{m}{s} \, °C \right] \qquad (9.5)$$

The convective term can be neglected, as the time scale for the upper layer of the Baltic is about 10 years (the time scale used here is the volume divided by the total volume flux).

The heat flux can be converted into a buoyancy flux by multiplying by the thermal expansion coefficient α, and the acceleration of gravity g

$$g(\overline{\Delta' v_3'})_w = \alpha(\overline{\theta' v_3'})_w \, g \qquad (9.6)$$

In the actual temperature range α is subject to a great variation. A functional approximation to the dependence is

$$\alpha \sim 1.35 \times 10^{-5} (\theta - 2) \qquad 2^{\circ}C < \theta < 15^{\circ}C \qquad (9.7)$$

which yields an average buoyancy flux of

$$o\left(g(\overline{\Delta'v'_3})_w\right) \simeq 3 \times 10^{-9} (\theta - 2) \ [m^2/s^3 \ ; \ \theta \ in \ ^{\circ}C] \qquad (9.8)$$

Exercise 9.1

Show, by using the findings in example 2.4 that the rate
of entrainment due to free penetrative convection in the Bornholm Basin (Fig. 9.3 b) amounts to half a meter per month on
average when $\theta \simeq 10^{\circ}C$.

The combined effects of free penetrative convection and
wind generated erosion in an estuary like the Baltic are discussed in Ch. 10.

The physics of the free penetrative convection may be described as follows (see Fig. 9.4). The uniform heating (or cooling) of the wall creates a heat flux which in the molecular diffusion layer is transported by molecular action (Brownian movements). In the boundary transition layer the heated fluid con-
verges along lines and then plunges upwards in sheets - a highly
intermittent process. These buoyant sheets are nearly two-dimensional plumes (see Ch. 12). Accordingly, the lower boundary
condition for free penetrative convection can be thought of as
an infinite number of plumes flowing into the convective layer
at an intermittent rate and from a line statistically distributed
over the area. This interpretation is useful because, as shown
in Ch. 12 on plumes, two-dimensional plumes are characterized
by having a constant velocity, independent of the distance from
the outlet. Therefore, we may approximate the total turbulent
convective layer as a region of "turbulent" plumes, where the
velocity i.e. the turbulent kinetic energy

$$\bar{e} = \tfrac{1}{2} \ \rho \overline{u'_i u'_i} \qquad (9.9)$$

is constant.

Hence, we take the turbulent kinetic energy $\bar{e}$ as nearly
constant and the buoyancy fluxes as linearly distributed (see
Fig. 9.5). The downward flux stems from the entrainment, which

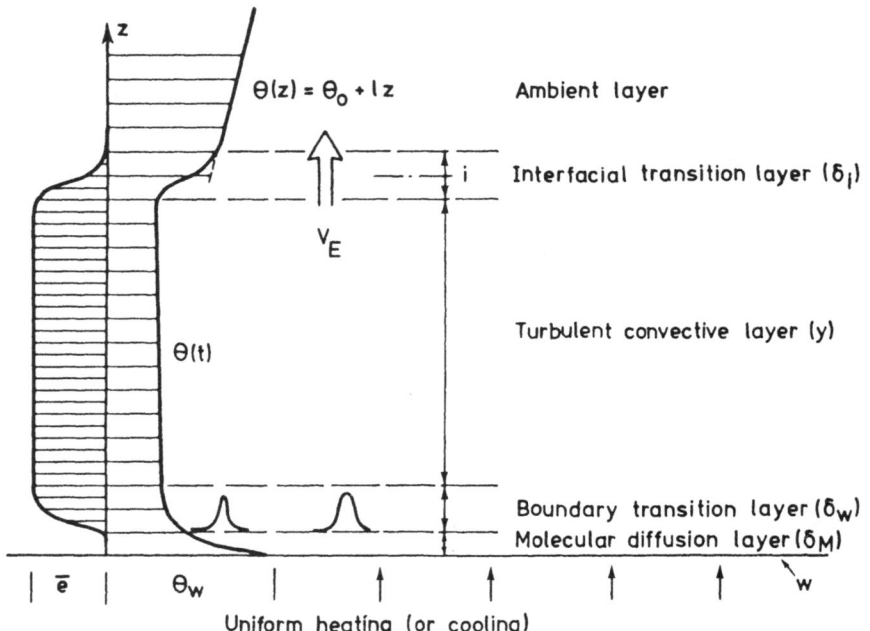

Fig. 9.4   Sketch (not in scale) of the different zones in a free penetrative convection process. θ is temperature (potential), ē is turbulent kinetic energy = ½ ρ $\overline{u'_i u'_i}$, w ~ wall, i ~ inversion or interface, ℓ is the temperature lapse rate in the ambient layer.

is initiated by the domes and streamers formed at the interface by the bombardment by the rising plumes.

As no mean flow is present, the only relevant equation of motion is the turbulent kinetic energy equation (Ch. 4) which degenerates into

$$\frac{\partial \overline{e}}{\partial t} = - \rho g \alpha \, \overline{v'_z \theta'} - \frac{\partial}{\partial z} (\overline{p' v'_z} + \overline{v'_z e}) + \rho \epsilon \tag{9.10}$$

where α is the volume coefficient of thermal expansion (a function of θ and hence of time) and θ the potential temperature.

If vertically propagating gravity waves in the stable fluid above are neglected, the turbulent kinetic energy equation per unit area integrated over the mixed layer depth yields

$$\int_0^Y \frac{\partial \overline{e}}{\partial t} \, dz = \int_0^Y - \rho g \alpha \, \overline{v'_z \theta'} \, dz + \int_0^Y (\rho \epsilon) dz \tag{9.11}$$

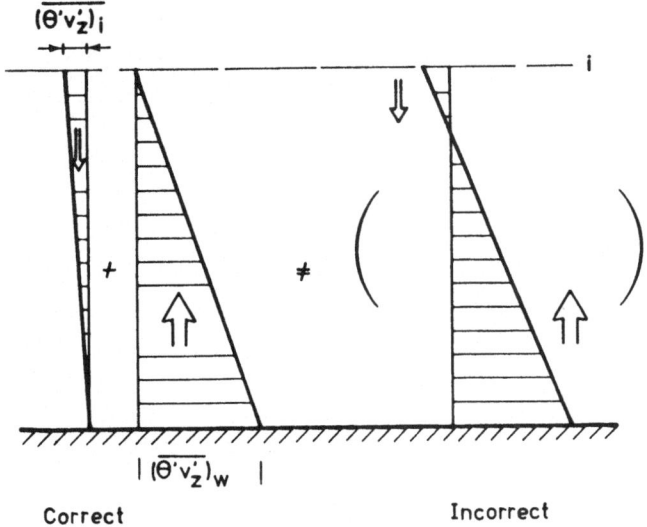

Fig. 9.5 The buoyancy fluxes in the well-mixed layer in free penetrative convection. Upward flux produces energy. Downward flux consumes energy.

The thickness of the well-mixed layer y is a function of time due to the rate of penetration of the mixed layer in the stable layer

$$V_E = \frac{\partial y}{\partial t} \tag{9.12}$$

This entrainment is the carrier of the downward heat flux

$$[\overline{v_z' \theta'}]_i = V_E \, \Delta \theta_i \tag{9.13}$$

Now the single terms in Eq. (9.11) can be evaluated:

$$\int_0^y \left(\frac{\partial \overline{e}}{\partial t}\right) dz = \frac{\partial}{\partial t}\left(\int_0^y \overline{e} \, dz\right) - \overline{e}_i \, \frac{\partial y}{\partial t} = \frac{\partial}{\partial t}\,(y\langle\overline{e}\rangle) - \overline{e}_i V_E \tag{9.14}$$

For convenience we introduce the velocity scale W_F, defined by the relation

$$W_F = \sqrt[3]{g \, \alpha \, y[\overline{v_z' \theta'}]_{source}} \tag{9.15}$$

107

where the radicant is proportional to the loss in poten-
tial energy of the buoyancy released (POT = $\frac{1}{2}$ g α y $[v_z'\theta']_{source}$)

The depth and time integrated turbulent kinetic energy is
related to W_F by

$$\langle \bar{e} \rangle = \frac{\rho W_F^2}{\sqrt{c_D}} \qquad (9.16)$$

where c_D = 20 (experimentally determined).

By introducing the velocity scale W_F and by assuming that
the ambient, stable layer fluid has no turbulent kinetic energy ($\bar{e}_i/\langle \bar{e} \rangle$ << 1), we obtain

$$\int_0^Y \left(\frac{\partial \bar{e}}{\partial t}\right) dz \simeq \frac{\rho \, W_F^3}{\sqrt{c_D}} \frac{V_E}{W_F} + y \frac{\partial}{\partial t} \left(\frac{\rho \, W_F^2}{\sqrt{c_D}}\right) \qquad (9.17)$$

for the rate of increase in turbulent kinetic energy per unit
area, where the last term is an order of magnitude less than the
first term (see Eq. (9.25)) - and hence quasi-stationarity is
assured.

The buoyancy fluxes must be divided into two parts, as
they - from a theoretical point of view - could be caused by
two different physical properties (for example heat/salt in
water). Therefore the integrated buoyancy fluxes give

$$\int_0^Y - \rho g \alpha \, \overline{v_z'\theta'} \, dz = \frac{1}{2} \, \rho g \alpha y [\overline{v_z'\theta'}]_0^Y = \frac{1}{2} \, \rho g \alpha y ([\overline{v_z'\theta'}]_w - V_E \, \Delta\bar{\theta}_i) \qquad (9.18)$$

where index i stands for the interface (or inversion base) and
index w for the wall. The fluxes ($[\overline{v_z'\theta'}]$) have been taken as positive values, and in accordance with the well-mixed stage, the
fluxes are quasi-linearly distributed over the depth (therefore
the factor $\frac{1}{2}$). The upward flux is the supplier of energy. The
downward flux is the energy absorber.

If we denote the dissipation

$$\int_0^Y (\rho\varepsilon) dz = - \text{DISS}$$

the turbulent kinetic energy equation for free penetrative convection states:

108

$$\frac{\rho W_F^3}{\sqrt{c_D}} \frac{V_E}{W_F} + y \frac{\partial}{\partial t}\left(\frac{\rho W_F^2}{\sqrt{c_D}}\right) = \frac{1}{2}\rho g \alpha y \left\{ [\overline{v_z'\theta'}]_w - V_E \Delta\overline{\theta}_i \right\} - DISS$$

$$(9.19)$$

As the main objective of outlining this equation is to evaluate the entrainment, the bulk flux Richardson number is introduced. Due to the non-stationary term the definition of $\mathbb{R}_f^T$ is slightly different from the ordinary definition, namely:

$$\mathbb{R}_f^T = \frac{\left(PROD - y \frac{\partial <\overline{e}>}{\partial t}\right) - DISS}{\left(PROD - y \frac{\partial <\overline{e}>}{\partial t}\right)} =$$

$$\frac{\frac{1}{2}\rho g \alpha y \,\overline{\Delta\theta}_i\, W_F + \frac{\rho W_F^3}{\sqrt{c_D}}}{\frac{1}{2}\rho g \alpha y \,[\overline{v_z'\theta'}]_w - y \frac{\partial}{\partial t}\left(\frac{\rho W_F^2}{\sqrt{c_D}}\right)} \frac{V_E}{W_F}$$

$$(9.20)$$

which may be taken equal to a constant, see Ch. 5:

$$\mathbb{R}_f^T = 0.18 \qquad \text{for } \mathbb{F}_\Delta^2 > \mathbb{F}_{\Delta,cr}^2$$

$$(9.21)$$

as plumes are the basic flow.

It is worth while noticing that the gain in potential energy due to entrainment is calculated in a different way from the calculations normally performed in the literature on the subject. It is usual to add the upward and downward fluxes and then calculate the gain as the small contribution from the resulting downward flux (see Fig. 9.5). This is of course wrong. (Just think of the two buoyancy fluxes as created by heat and salt, respectively - they would both be uniformly distributed over the depth).

Before we proceed with the calculation for obtaining an entrainment function we look at Eq. (9.20) in the limit of no buoyancy difference between the stable layer and the well-mixed convective layer, which then states

$$\frac{V_{E,0}}{W_F} = \mathbb{R}_f^T \frac{\frac{1}{2} \rho g \alpha y(t) [\overline{v_z' \theta'}]_w - y \frac{\partial}{\partial t}\left(\frac{\rho W_F^2}{\sqrt{c_D}}\right)}{\frac{\rho W_F^3}{\sqrt{c_D}}} = c_0 \qquad (9.22)$$

where c_0 is a constant which can be determined by Eq. (9.22). Let us first calculate the rate of change in the level of turbulent kinetic energy:

$$y \frac{\partial}{\partial t}\left(\frac{\rho W_F^2}{\sqrt{c_D}}\right) = \rho \frac{y}{\sqrt{c_D}} \frac{\partial W_F^2}{\partial y} \frac{\partial y}{\partial t} = \frac{\rho y}{\sqrt{c_D}}\left(\frac{2}{3} \frac{W_F^2}{y}\right)V_{E,0} = \frac{2c_0}{3\sqrt{c_D}} \rho W_F^3$$

$$(9.23)$$

If we insert this expression in Eq. (9.22) we obtain the following equation

$$c_0 = \mathbb{R}_f^T \frac{\frac{1}{2} \rho W_F^3 - \frac{2c_0}{3\sqrt{c_D}} \rho W_F^3}{\frac{\rho W_F^3}{\sqrt{c_D}}} \qquad (9.24)$$

which yields the constant $c_0 = 0.36$.

With known constants we are able to estimate the order of magnitude of the rate of change of the turbulent kinetic energy, which yields

$$0\left(y \frac{\partial}{\partial t}\left(\frac{\rho W_F^2}{\sqrt{c_D}}\right) \Big/ \left(\frac{\rho W_F^3}{2}\right)\right) \sim 0.1 \qquad (9.25)$$

in accordance with the findings of Deardorff, Willis and Lilly [1969].

Now let us make Eq. (9.20) dimensionless by introducing the characteristic velocity W_F as defined by Eq. (9.15):

$$\mathbb{R}_f^T = \frac{\frac{\rho W_F^3}{2} \frac{(\overline{v_z' \theta'})_i}{(\overline{v_z' \theta'})_w} + \frac{\rho W_F^3}{\sqrt{c_D}} \times \frac{V_E}{W_F}}{\frac{\rho W_F^3}{2} - \left(\frac{\rho W_F^3}{\sqrt{c_D}}\right)\frac{2}{3}\frac{V_E}{W_F}} \qquad (9.26)$$

This equation is, with deference to the historical traditions, solved with respect to the ratio of the downward to the upward heat fluxes, respectively:

$$\frac{(\overline{v_z' \theta'})_i}{(\overline{v_z' \theta'})_w} = \mathbb{R}_f^T - 0.50 \frac{V_E}{W_F} = \mathbb{R}_f^T \left[1 - \frac{V_E}{V_{E,0}} \right] \tag{9.27}$$

where $\mathbb{R}_f^T = 0.18$.

By inspection of Eq. (9.20) it is seen that the assumption of a constant ratio corresponds to a constant ordinary flux Richardson number if the gain in turbulent kinetic energy of the entrained fluid is neglected. The reason for the success of assuming a constant ratio in parameterization schemes is the fact that despite the very start of the inversion process the gain in the turbulent kinetic energy plays a minor role.

If the total inversion rise process shall be parameterized under weak to neutral conditions in the air above, we have to take the gain in the turbulent kinetic energy as well as the gain in potential energy into account, which means that we shall apply Eq. (9.27), but it is not convenient when these more general cases are to be treated. Instead, we solve Eq. (9.20) with respect to the entrainment velocity with the following results:

$$\frac{V_E}{W_F} = \frac{\mathbb{R}_f^T \ \mathbb{F}_{\Delta,w}^2}{1 + 0.5 \ \mathbb{F}_{\Delta,w}^2} \quad ; \quad \mathbb{R}_f^T = 0.18 \tag{9.28}$$

where we have defined the densimetric Froude number squared as

$$\mathbb{F}_{\Delta,w}^2 = \frac{W_F^2}{gy \ \alpha \ \overline{\Delta\theta}_i} \tag{9.29}$$

Eq. (9.28) shows the same behaviour as other entrainment functions, i.e. proportionality to the densimetric Froude number squared for small values of $\mathbb{F}_{\Delta,w}^2$, while the entrainment goes to a constant for infinite value of $\mathbb{F}_{\Delta,w}$. This entrainment function is shown in Fig. 9.6. The deviation due to the turbulent kinetic energy starts at above approximately $\mathbb{F}_{\Delta,w}^2 = 0.1$, which therefore may be used as a practical limit to the commonly used assumption of a constant flux ratio. For increasing values of $\mathbb{F}_{\Delta,w}$ the rate of entrainment increases more slowly and finally reaches

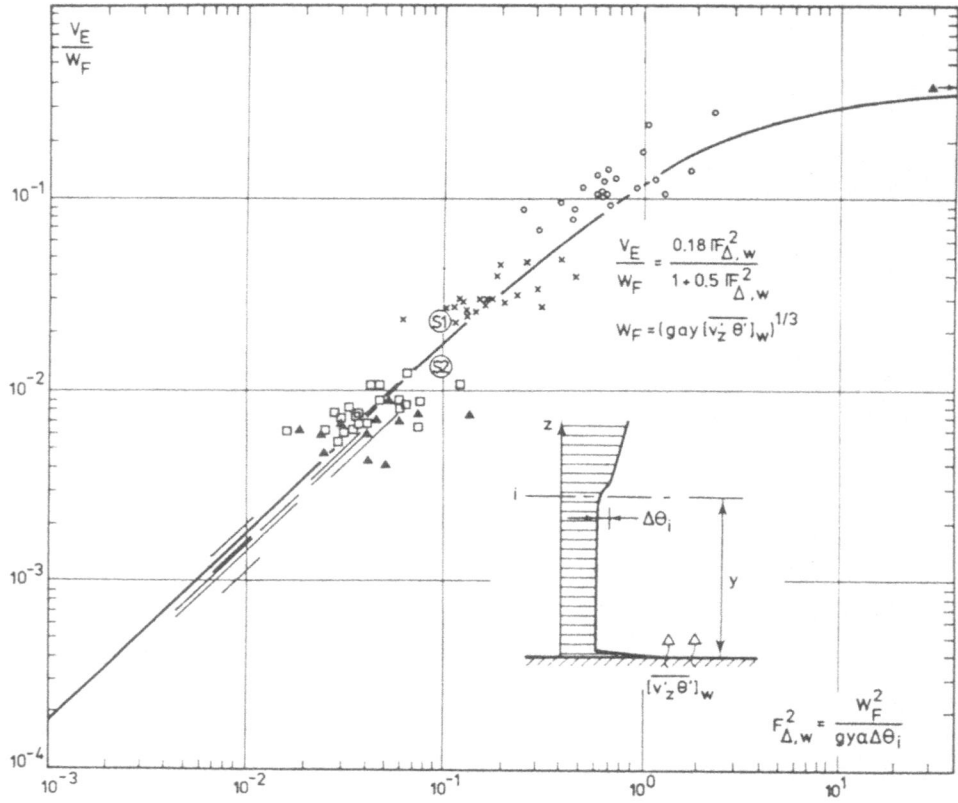

Fig. 9.6   The entrainment function for free penetrative convection.

The points are based on data referred by
o, x, □   *Heidt [1975] (Laboratory experiments.*
S1, S2   *Willis and Deardorff [1974] (Laboratory experiments).*
▲   *Farmer [1975] (Field data from solar heating beneath lake ice).*
/   *Bo Pedersen and Jürgensen [1984] (Laboratory experiments).*

the value of $0.36 \, W_F$ in accordance with the measurements by Farmer [1975].

In Part III, an experimental set-up for entrainment measurements in free convection is illustrated, Bo Pedersen and Jürgensen [1984].

10. WINDDRIVEN STRATIFIED FLOW

The wind is often one of the most important external forcing functions with respect to the currents and mixing processes in the ocean, in estuaries and lakes.

Let us by one example illustrate some of the fundamental processes occurring in a shallow semi-enclosed stratified body of water located on the northern hemisphere, see Fig. 10.1.

We have indicated the external forces of major importance for the stratified body of water, but in this example we shall be concerned mainly with the combined effects of the wind and of the heating/cooling.

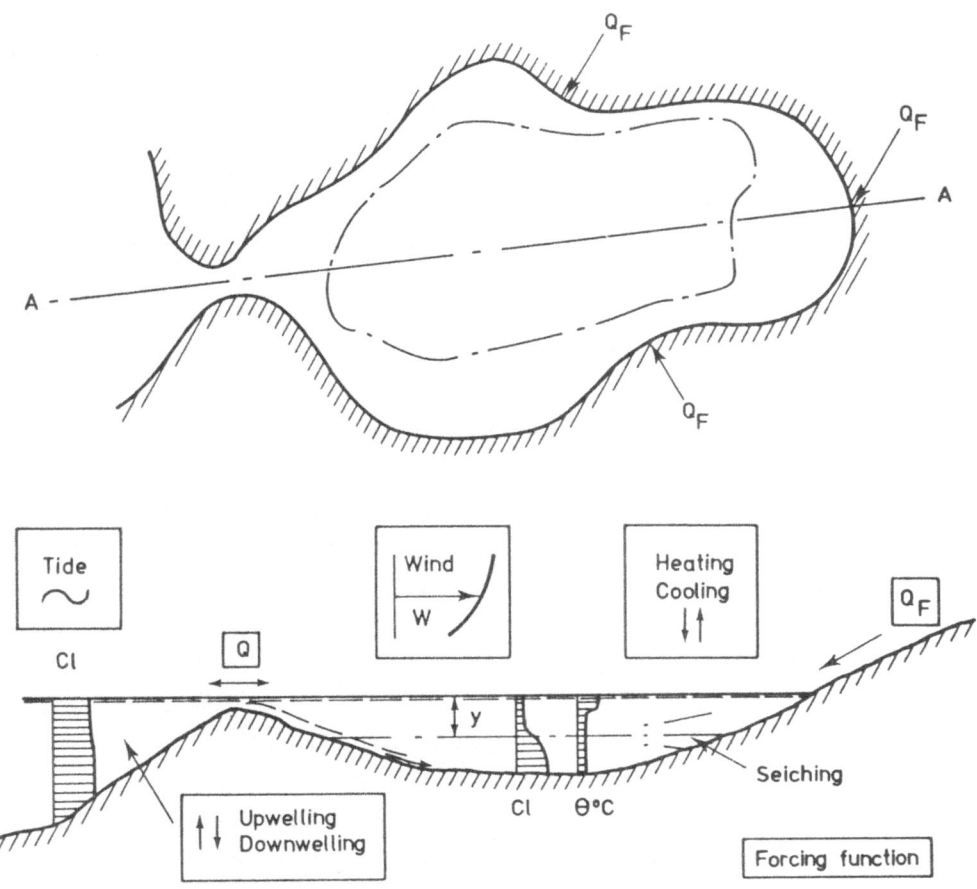

Fig. 10.1 *Shallow semi-enclosed density stratified body of water on the northern hemisphere (distorted scale).*

When a wind field is imposed on a stratified body of water, the first response is an acceleration of the near-surface water. As time goes, the depth of the flowing water increases, the rate of increase being determined by the dynamic balance between the rate of acceleration and the shear stress (see example 4.2.3). This initial deepening represents the highest possible rate of entrainment, and it soon causes the flow to reach the interface which may be either a thermo- or a halocline. The increased stability slows down the entrainment velocity, which normally decreases by several orders of magnitude when the interface is reached. If the wind is persistent and the water body of limited extension, the boundary effects soon become an important factor. The surface water, initially transported in the direction of the wind, starts a set-up in the downstream part of the water body, causing an increased pressure in the surface layer and, accordingly, a set-down of the interface. This piling-up of light water in the downstream direction continues until a nearly steady state is eventually reached in which the dynamic balance is between the surface slope induced pressure gradient and the wind imposed shear stress (see Ch. 3.3.2). We have sketched this fully developed flow situation in Fig. 10.2. The flow pattern is normally rather complicated with average upper layer flows in the downwind as well as in the upwind direction and with vertical velocity vectors turning clockwise as well as counterclockwise with depth - all due to the boundary effects. This has for example, been demonstrated by measurements in the smaller Swedish lakes, see Bengtsson [1978]. A theoretical approach to this shape factor effect has been given by Engelund [1973], see Part III.

When the body of water (Fig. 10.1) has a large horizontal extension, the effect of the earth's rotation becomes a significant factor in the flow development process. In a rotating system far from the shore, the velocity vector turns to the right (on the northern hemisphere) and decreases with depth. In a homogeneous fluid this so-called Ekman spiral has a vertical extension of the Ekman depth, determined by the dynamic balance between the imposed wind shear stress at the water surface and the Coriolis force of the water flow which by dimensional reasoning, yields

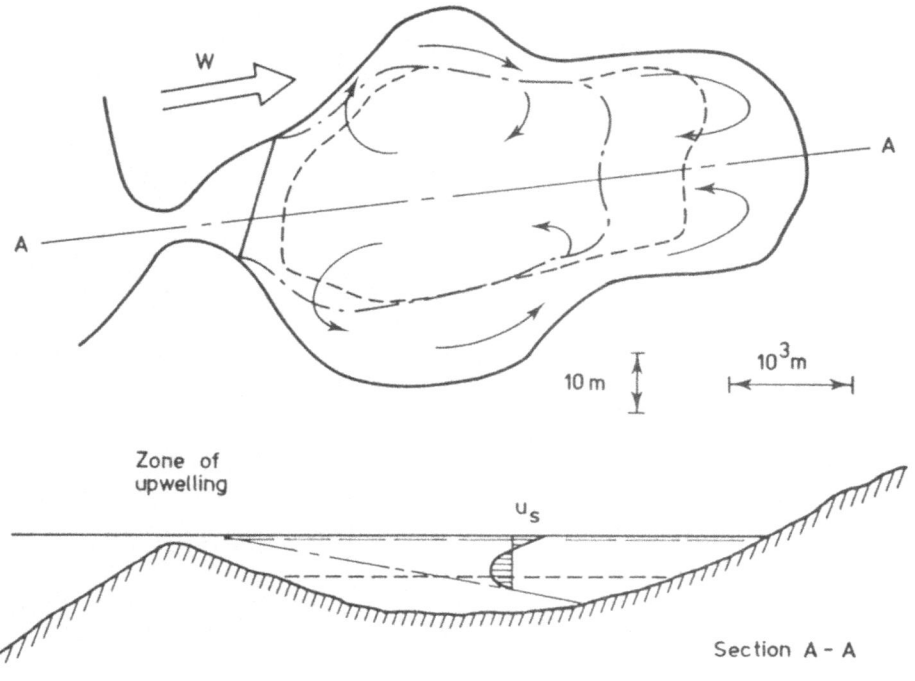

Fig. 10.2   *Initial and final position of the interface in*
in a small stratified body of water suddenly
loaded by a persistent wind. The average upper
layer velocities are indicated.

$$L_{Ekman} = \frac{U_F}{f_c} \qquad\qquad (10.1)$$

where

$U_F = \sqrt{\dfrac{\tau_w}{\rho}}$ = the friction velocity at the water surface

$f_c = 2\omega \sin\phi$ = the Coriolis parameter.

At the coastlines in the wind direction the turning to the right of the velocity vector creates up- or downwelling of the denser water underlying, associated with large vertical movements of the interface, and hence creating large circulations in the adjacent fjords.

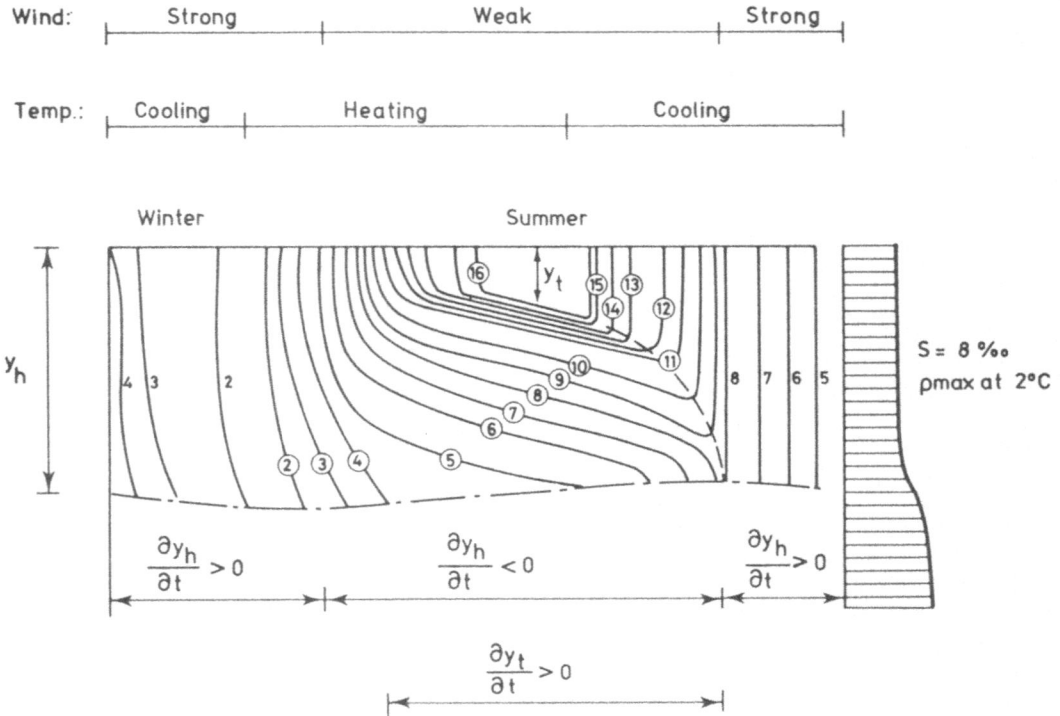

Fig. 10.3
Fig. 10.3 An example of an annual cycle of wind, heating and cooling in an estuary with a continuous supply to the deep water. y_h = halocline depth, y_t = thermocline depth, numbers in the figure are °C.

In the nearly steady state a large amount of potential energy has been accumulated in the system. When the wind load on the system is removed, a seiching therefore starts in the upper and in the lower layer, producing turbulence and hence mixing. An example on seiching in a fjord is given in Part III, Møller and Bo Pedersen [1983].

Heating or cooling of the upper part of a surface layer may generate a buoyancy flux at the surface. Let us, for example, look at the effect on the temperature field in the water body, Fig. 10.1, of the annual heating and cooling cycle, see Fig. 10.3, where the annual variation of the wind field has been indicated as well.

As the salinity of the present example is approximately eight per thousand, the water has its maximum density at around

two degrees centigrade. Consequently, the buoyancy effects aris-
ing from the heating/cooling cycle can be described as follows:

When the heating begins, the temperature is below the
maximum density temperature, and therefore the heating creates
an unstable surface layer which gives rise to a free penetrative convection, see Ch. 9, with maximum penetration velocity
through the upper homogeneous layer, and with a radically reduced penetration when it reaches the stable halocline. This
process continues until the whole water column has reached the
temperature of maximum density (2 ^{o}C in the present example).
Further heating now creates a stable upper layer, but due to
the strong wind, the energy input into the system is strong
enough to hinder the formation of a thermocline. The increasing
rate of heating and the decreasing meteorological activity may
finally start the thermocline formation. In the example illustrated by Fig. 10.3, the energy input caused by the wind is
strong enough to erode the thermocline, at least on a monthly
average. When looking at the behaviour on a smaller time scale
we may find the picture quite different, due to the delicate
balance between the stabilizing heating and the eroding wind.

When the cooling period starts, an unstable surface zone
is formed, and consequently we observe a fast downward free penetration into the layer above the stable thermocline. This process combined with the wind mixing yields an increasing rate of
erosion of the thermocline (see the dashed line in Fig. 10.3).
When the strong winds begin, we again obtain a nearly homogeneous upper layer which continues to exist - with decreasing
temperature and, later, increasing temperature - as long as the
strong winds prevail.

The position of the halocline in the present example (the
Bornholm Basin) is determined by the balance between the nearly
continuous supply of dense deep water and the wind generated
erosion. This yields, as indicated in Fig. 10.3, a rise of the
halocline during the period of weak winds (and high stability
above) and a lowering during the period with high meteorological activity.

With focus on the entrainment process, the most characteristic feature of winddriven flows is the fact that most of the

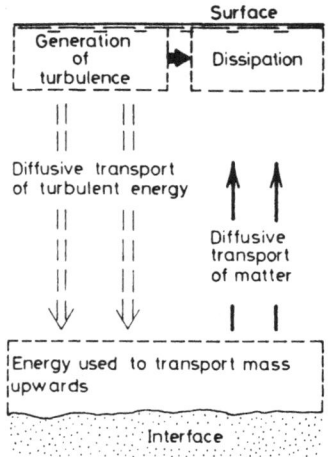

Fig. 10.4 Energy exchange in wind-driven flow. (From Ottesen Hansen [1975].

production takes place near the water surface, see Fig. 10.4 (from Ottesen Hansen [1975]). A great part of this surface production is dissipated in the surface layer as well, the rest is transported downwards by turbulent diffusion and is partly dissipated and partly used for diffusive transport against gravity of dense water.

Now, let us return to the dynamic description of wind driven stratified flows. In the initial phase just after the wind has been imposed on the water surface, the dynamic balance is simply between the local rate of acceleration and the imposed shear stress (example 4.2.3),

$$\frac{\partial u}{\partial t} = \frac{\partial}{\partial x_3} \left(\frac{\tau}{\rho}\right) \tag{10.2}$$

If we make an acceptable assumption as to the eddy viscosity, Eq. (10.2) can be solve, see for instance Ottesen Hansen [1975]. Denoting the friction velocity at the surface

$$U_F = \sqrt{\tau/\rho} \tag{10.3}$$

we find an initial rate of erosion which is

$$\frac{V_E}{U_F} = o(0.1) \tag{10.4}$$

Therefore, the order of magnitude of the time needed to reach the interface is

$$o(T) \simeq \frac{y}{V_E} = \frac{y}{V_E} \frac{U_F}{U_F}$$

which, for a typical depth of approximately 10 m and a wind velocity of approximately 10 m/s, yields ($U_F \simeq 10^{-2}$ m/s)

$$T \simeq o\left(\frac{10}{0.1} \frac{1}{10^{-2}}\right) \simeq o(10^4 \text{ s})$$

or some hours. During so short a period the boundaries cannot normally have influenced the overall flow (U_s T ~ a few kilometres).

If the shore boundaries have not affected the flow significantly in the initial phase, the next phase is characterized by a decreased downward rate of erosion. If we look at the integrated momentum equation for the upper layer, we have at this stage:

$$\frac{d}{dt} (\rho y V) = \tau_s - \tau_i \tag{10.5}$$

which shows that the further development is highly dependent on the extent to which the lower layer is able to balance the shear stress. Initially there are two possibilities, namely entrainment ($\tau \sim \rho V_E u_i$) and acceleration of the lower fluid. In this initial phase of the lower layer the rate of entrainment is determined by the production of turbulent kinetic energy in the upper layer (by the bulk flux Richardson number), while the remaining part of the imposed interfacial shear stress creates an acceleration of the lower layer fluid. The continued increase in velocity of the upper layer increases the interfacial shear stress, which, in turn, decreases the rate of acceleration and the rate of production (a greater part of the imposed energy being transported to the lower layer ~ $u_i \tau_i$). Hence, we may observe a quasi-stationary condition in this second phase.

The second phase is gradually changed due to the piling-up of light water downstream, which builds up a pressure gradient to counter-balance the shear stress and, accordingly, gradually to take over the role of the non-stationary term. As mentioned above, the flow pattern and hence the production are rather complicated in this phase due to the influence of the boundaries.

The general way in which to calculate the rate of entrainment in wind driven flow is by making use of the bulk flux Richardson number concept, Ch. 5, according to which we have

$$\mathbb{R}_f^T = \frac{V_E[\frac{1}{2} \xi \rho \Delta gy + (\rho \delta <\bar{e}> - \bar{e}_i)]}{PROD - q \frac{\partial}{\partial s}(\rho \delta_e <\bar{e}>)} \tag{10.6}$$

The convective term may be neglected in wind driven flows in nature. Consequently, by introducing the variable c_e by

$$\delta <\bar{e}> - \bar{e}_i = \frac{1}{2} c_e U_F^2 \tag{10.7}$$

and taking the upper layer as being homogeneous ($\xi = 1$) we can write

$$\mathbb{R}_f^T = \frac{V_E \frac{1}{2} \rho \Delta gy(1 + c_e \mathbb{R}i_F^{-1})}{PROD} \tag{10.8}$$

where $\mathbb{R}i_F$ is defined below.

As we are normally faced with a complicated flow pattern, a general calculation of the production term is not straightforward.

In the following we shall make some crude approximations in order to evaluate the PROD-term and thus to obtain an entrainment function. By definition (Ch. 4) we have

$$PROD = -\int_0^y \left(\tau \frac{\partial u}{\partial x_3}\right)dx_3 = \int_0^y \left(u \frac{\partial \tau}{\partial x_3}\right)dx_3 - [u\tau]_{x_3=0}^{x_3=y} \tag{10.9}$$

where the right-hand side is often more convenient to evaluate. A linear shear stress distribution, for example, yields

$$PROD \simeq -(\tau_s - \tau_i)V + U_s \tau_s - u_i \tau_i = \tau_s(U_s - V) + \tau_i(V - u_i) \tag{10.10}$$

In order to proceed further we take

$$\frac{\tau_s}{\rho} \simeq \frac{f_s}{2} (U_s - V)^2 \qquad U_F = \sqrt{\frac{\tau_s}{\rho}} \qquad\qquad (10.11)$$

$$\frac{\tau_i}{\rho} = \frac{f_i}{2} (V - u_i)^2 \qquad\qquad\qquad\qquad (10.12)$$

which yields

$$PROD \simeq \rho \sqrt{\frac{2}{f_s}} \, U_F^3 \left(1 + \left(\frac{\tau_i}{\tau_s}\right)^{3/2} \left(\frac{f_s/2}{f_i/2}\right)^{1/2} \right) \qquad (10.13)$$

In the quasi-stationary wind driven flow the interfacial shear stress is small compared with the surface shear stress (see for example Ottesen Hansen [1975]), and therefore, we can evaluate the production as

$$PROD \simeq \rho \sqrt{\frac{2}{f_s}} \, U_F^3 \qquad\qquad\qquad (10.14)$$

We obtain this expression for the production of turbulence per unit area in the upper layer by using some assumptions, the integrated validity of which it may be difficult to estimate. Therefore it is worthwhile to mention that the energy input from the shear stress to the surface drift current U_s is

$$Input = U_s \, \tau_s \simeq \left(\sqrt{\frac{2}{f_c}} \, U_F \right) \rho \, U_F^2 \qquad\qquad (10.15)$$

i.e. corresponding to the production evaluated above. This is in agreement with the observations in the field, which show that the wind-generated mixing causes an upwards directed entrainment, from the non-turbulent (hardly any production, as $u_i \tau_i \ll U_s \tau_s$) to the highly turbulent region. Furthermore, it has been verified directly by, for example, Kullenberg [1976].

In the cases where we may estimate the production as the energy input at the water surface we consequently have en entrainment function, which reads

$$\frac{V_E}{U_F} = \frac{2 \, \mathbb{R}_f^T}{\mathbb{R}i_F} \frac{\sqrt{2/f_s}}{(1 + (c_e/\mathbb{R}i_F))} = \frac{2 \, \mathbb{R}_f^T \sqrt{2/f_s}}{c_e + \mathbb{R}i_F} \qquad (10.16)$$

where the bulk Richardson number commonly used has been introduced

121

$$\text{IRi}_F = \frac{\Delta g y}{U_F^{\ 2}} \tag{10.17}$$

In the field we can take the friction factor to be equal to a constant when the flow is fully developed, see Ch. 7. For developing flows, i.e. in the initial phase where the erosion takes place through the nearly homogeneous layer ($\text{IRi}_F < 1$) the friction factor is more likely to be an order of magnitude higher, see for instance Bo Pedersen [1972]. We may therefore estimate the numerator in Eq. (10.16) as

$$2 \ \text{IR}_f^T \sqrt{2/f}_s \simeq \begin{cases} 2 \times 0.045 \sqrt{1/(1.5 \times 10^{-3})} = 2.3 & \text{IRi}_F > \text{IRi}_{F,cr} \\[2ex] 2 \times 0.18 \ \sqrt{1/10^{-2}} \quad\quad = 3.6 & \text{IRi}_F < \text{IRi}_{F,cr} \end{cases} \tag{10.18}$$

where we have used the same values for the bulk flux Richardson number as in all the other entrainment functions.
The term

$$c_e = 2 \ \frac{\delta <\overline{e}> - e_i}{U_F^{\ 2}} \tag{10.19}$$

has a bearing on the entrainment function only, when IRi_F is small, i.e. in the developing, supercritical flow.

We may estimate this term to be $c_e \sim 6$, based on the general expression from turbulence models

$$\overline{e} = c_D^{-\frac{1}{2}} \ U_F^{\ 2} \simeq 3.5 \ U_F^{\ 2} \tag{10.20}$$

We shall not be further concerned with this term, as it has no practical importance, since the initial rate of erosion is artificially defined, see Ottesen Hansen [1975], and the value $c_e \simeq 6$ yields a reasonable entrainment function.

The uncertainty in the numerator for the low values of IRi_F combined with the small difference between the numerators for super- and subcritical flows makes it reasonable to operate with a single entrainment function

$$\frac{V_E}{U_F} \simeq \frac{2.3}{6 + \mathbb{R}i_F} \tag{10.21}$$

where we have chosen the numerator valid for high values of $\mathbb{R}i_F$, this being by far the most common one in nature.

The entrainment function equation 10.21 is only an example of an entrainment function, valid for the flow situations, in which we may estimate the total energy input at the surface to be consumed in the upper well-mixed layer.

We have plotted the entrainment function Eq. (10.21) in Fig. 10.5. The observed values of the entrainment stem from the laboratory as well as from the field. Field measurements always suffer from considerable uncertainties, which is reflected in the large scattering. Part of this apparent scattering may be caused by the special conditions of the entrainment process, such as smallness of the lakes, which make only a fraction of the energy input available for the mixing process as pointed out by Ottesen Hansen [1978].

It is emphasized that the entrainment function presented is not general. In complicated cases, such as those normally present in small lakes, it is recommended to use the bulk flux Richardson number concept directly, preferably in connection with a numerical model.

Example 10.1

In exercise 2.2 the wind generated erosion of a two-layered system was treated in an approximative way.

Let us make a more accurate estimate of the time needed to mix a water column of $y_1 = 10$ m (upper layer), $y_2 = 20$ m (lower layer), initial density difference $\Delta = 10^{-3}$ due to a wind $w = 10$ m/s with a friction coefficient of $f/2 = 2.6 \times 10^{-3}$.

The conservation of mass (Eq. (4.1.5)) states

$$\frac{\partial}{\partial t} (\Delta y_1) = 0 \tag{10.22}$$

and hence,

$$\Delta y_1 = \Delta_0 y_0 = 10^{-3} \times 10 = 10^{-2} \text{ [m]} \tag{10.23}$$

(where index 0 stands for initial values).

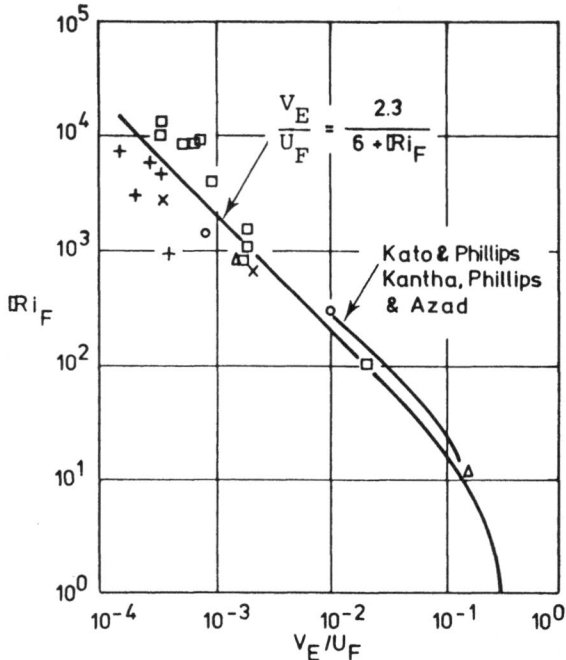

$$\frac{V_E}{U_F} = \frac{2.3}{6 + Ri_F}$$

Kato & Phillips
Kantha, Phillips & Azad

Fig. 10.5 An example of one entrainment function for winddriven flow (see text).
Observations:
Kato and Phillips [1969]
Kantha, Phillips and Azed [1977]

□ *The sea*
△ *Lake Ontario* } *Kullenberg [1977]*

o *Lake Windermere*
x *Lake Esrum* } *Ottesen Hansen [1975]*
+ *Lake Velen*

This implies that the Richardson number for the mixing situation is a constant

$$Ri_F = \frac{\Delta g y}{U_F^2} = \frac{10^{-2} \times 9.81}{3.25 \times 10^{-4}} = 302 \qquad (10.24)$$

where

$$U_F = \sqrt{\frac{\tau}{\rho_w}} = \sqrt{\frac{2.6 \times 10^{-3} \times 10^2}{800}} = 1.80 \times 10^{-2} \ [m/s] \qquad (10.25)$$

has been introduced.

The continuity equation simply states

$$\frac{\partial y}{\partial t} = V_E = \frac{2.3\ U_F}{6 + \mathbb{R}i_F} = \frac{2.3 \times 1.80 \times 10^{-2}}{6 + 302} = 1.34 \times 10^{-4}\ [m/s] \tag{10.26}$$

and hence, the time lapsed for total mixing to the bottom

$$T = \frac{y_2}{V_E} = \frac{20}{1.34 \times 10^{-4}} = 1.5 \times 10^5\ s \sim 1.7\ days \tag{10.27}$$

(where the estimate in Exercise 2.2 gave 2.2 days).

Example 10.2

In example 4.1.2, a wind-mixed fjord was treated. Let us make an estimate of the increase in the discharge in a typical fjord of length L = 100 km, width B = 1 km, upper layer depth y $\sim$ 10 m and a fresh water discharge R = 500 m^3/s in the upstream part of the fjord.

The constant reduced mass flux is (Eq. (4.1.9)

$$\Delta Q = \Delta_0 R \approx 0.027 \times 500 = 13.5\ m^3/s \tag{10.28}$$

The continuity equation states (Eq. (4.1.6)

$$\frac{dQ}{dx} = V_E B = \frac{2.3\ U_F^3 BL}{gy\ \Delta_0\ R}\ Q\left(\frac{x}{L}\right) \tag{10.29}$$

where the entrainment function, Eq. (10.21), has been applied ($\mathbb{R}i_F \gg 6$). Eq. (10.29) is readily solved to yield

$$\frac{Q}{R} = \exp\left\{\frac{2.3\ U_F^3\ BL}{gy\ \Delta_0\ R}\left(\frac{x}{L}\right)\right\} \tag{10.30}$$

or with the actual figures (U_F as in Example 10.1)

$$Q = 500\ \exp\left\{\frac{x}{L}\right\} \tag{10.31}$$

which yields a discharge at the mouth of

$$Q_{max} = 1360\ m^3/s \tag{10.32}$$

i.e. increased by nearly a factor three.

A further example is given in Part III, Bo Pedersen and Møller [1981].

11. HORIZONTAL BUOYANT FLOW

A horizontal buoyant flow is the flow created by a horizontally directed source of mass, momentum, and buoyancy into an ambient fluid. The horizontal direction is preserved by the existence of two stable density jumps embedding the flow, as for instance a free surface and an interface.

We can divide the horizontal buoyant flows (see Fig. 11.1) in *overflows*, by which we mean flows in contact with the atmosphere, and *interflows*, which take place along a pycnocline in the interior of the ambient fluid. Within both types of flow we have supercritical flows, which are primarily momentum driven, and subcritical flows, which are primarily buoyancy driven.

The subcritical overflow is by far the most common flow within the geophysical field, as it is present in the oceans, in the estuaries as well as in some lakes and reservoirs. Two of the most important overflows in the Atlantic Ocean are the Gulf Stream, which carries an excess of heat from the equatorial region to the northern part of the ocean, and the Norwegian Coastal Current which originates from the precipitation over the Baltic watershed and Norway (see Fig. 4, Part III, Bo Pedersen [1980b], which shows a chart of the north-eastern part of the Atlantic Ocean). Dynamically, horizontal buoyant flows are very delicate as they are highly sensitive to any external forces, and they probably constitute the most difficult density stratified flow to treat, which is illustrated below.

Fiords

A brief review on fiords has been given by the author [1978 a]. The main parameters affecting the dynamics of fiords are illustrated in Fig. 11.2 and can be summarized as follows:

i)   The geometry.
ii) The hydrology of the adjacent watershed.
iii) The oceanographic conditions outside the fiord.
iv) The wind field.

Overflows

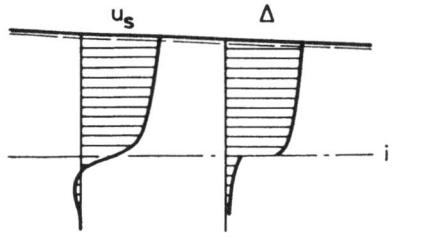 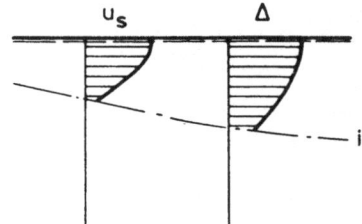

$\mathbb{F}_{\Delta} < \mathbb{F}_{\Delta,cr}$: Subcritical $\mathbb{F}_{\Delta} > \mathbf{F}_{\Delta,cr}$: Supercritical

Interflows

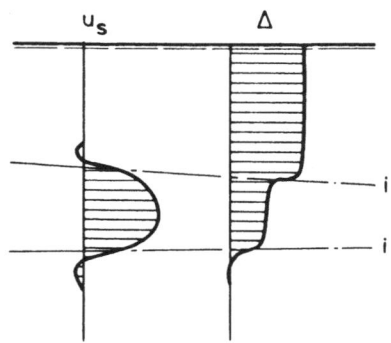 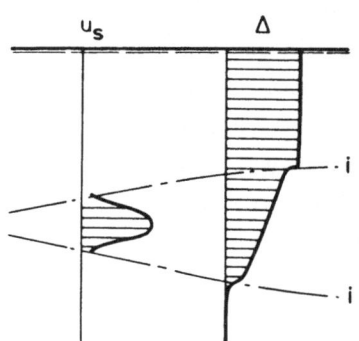

*Fig. 11.1   Definition sketch of overflows and interflows,
both subdivided into the subcritical and the
supercritical flow ranges, respectively.*

It proves quite natural to treat the upper and the lower
parts of the fiord separately (the lower part mentioned in Ch.8
on dense bottom currents). The integrated effects of the above-
mentioned parameters on the upper part of the fiord can be seen
from the salinity profiles, as demonstrated by Pickard [1961],
see Fig. 11.3. Types 1 and 2 can be characterized as the dynamically active fiords in the sense that the freshwater supply
is sufficient to create a circulation mode in the upper strata.
Type 3 fiords are in the same sense dynamically passive, only
reflecting the boundary conditions in the adjacent ocean - they
are merely a bay to the ocean.

All other parameters kept unchanged, a fiord has a tendency to change from type 1 to 2 and 3, when:

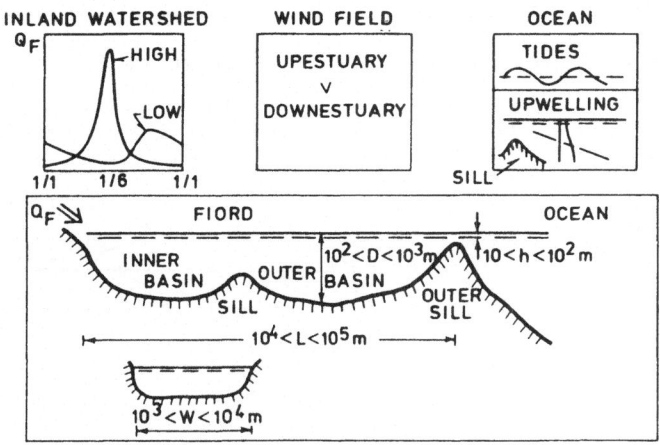

Fig. 11.2   The main parameters affecting the dynamics of fiords (from Bo Pedersen [1978 a].

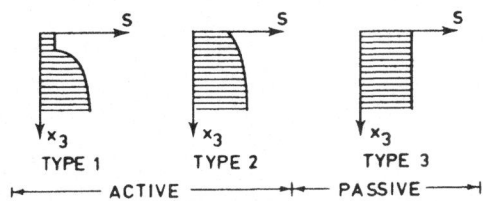

Fig. 11.3   Types of fiords. (After Pickard [1961].

Geometry: Length/outlet depth ratio increases (i.e. the tidally generated velocity and thus the mixing increases).

Hydrology: Freshwater discharge decreases (i.e. the critical depth and thus the layer depth decrease).

Ocean conditions: Tidal amplitude increases (i.e. the tidally generated velocities and thus the mixing increases).

Windfield: Wind velocity increases (i.e. the mixing increases).

As we shall mainly be concerned with the entrainment processes, we have in Fig. 11.4 illustrated the tidally induced variation in the longitudinal chlorinity distribution in the Alberni Inlet (from Tully [1949]). This demonstrates that although

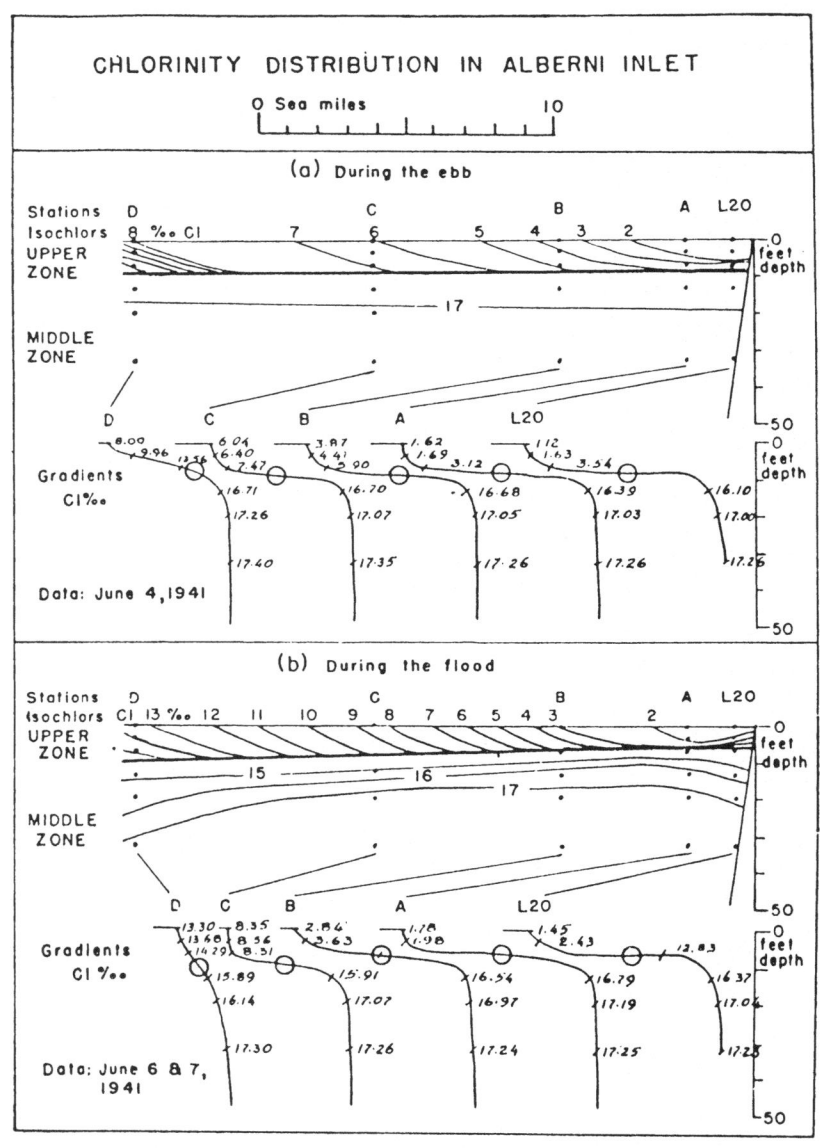

Fig. 11.4 Typical chlorinity distribution in Alberni Inlet (from Tully [1949]).

the tidal excursion is significant, the change in the dynamic head $\widetilde{\Delta y}$ is very small, simply because the longitudinal gradient of $\widetilde{\Delta y}$ is small. This is typical of the type 1 fjord with a high Δ-value and a nearly constant depth y, and it has a great bearing on the entrainment process, this dynamic head being part of the gain in potential energy for the entrained water, (POT = V_E $\frac{1}{2}$ $\xi \Delta \rho g y^2$). Type 1 fiords are characterized by having a tidally generated velocity not exceeding the order of magnitude of the freshwater-generated velocities (in Alberni Inlet they are of the same order of magnitude).

Neither the mean velocity nor the tidal velocity, but rather the wind is responsible for the turbulence productions creating entrainment in Alberni Inlet. The velocity and the density profiles, respectively, indicate that the type 1 fiords are associated with low values of the densimetric Froude number; they are subcritical flows.

In a type 2 fiord the pronounced interfacial jump in the density is missing, and the flow has the character of a super-critical density overflow, normally created by the combined effects of a weak freshwater discharge and a strong tidally generated flow. This is normally the case for the Norwegian fiords during the winter period. The flow behaving supercritically means that the dynamic head $\tilde{\Delta y}$ is small compared with the tidally generated velocity head ($v_{tidal}^2/2g$), which yields an average densimetric Froude number higher than the critical one. Accordingly, the entrainment associated is high, see below. The flow and the salinity pattern in a type 2 fiord are normally very complicated to describe in terms of physical causes, as they are very sensitive to the wind field and the boundary conditions which, in turn, both are very intermittent (especially during the winter time) due to the unstable meteorological conditions.

The Great Belt

The Danish National Agency of Environmental Protection started in 1973 a five years' intensive measuring programme (see Bæltprojektet [1976] in the Danish inland waters which connect the Baltic and the Atlantic. This programme involved measurements as well as numerical modelling (see DHI-Report [1977]) and was the Danish contribution to a more general investigation on the Baltic. The Danish inland waters, see Fig. 11.5 are interesting in many respects, first of all because they are the outer sill of the largest fiord in the world, and furthermore because they cover the whole spectrum of classes of estuaries, namely:

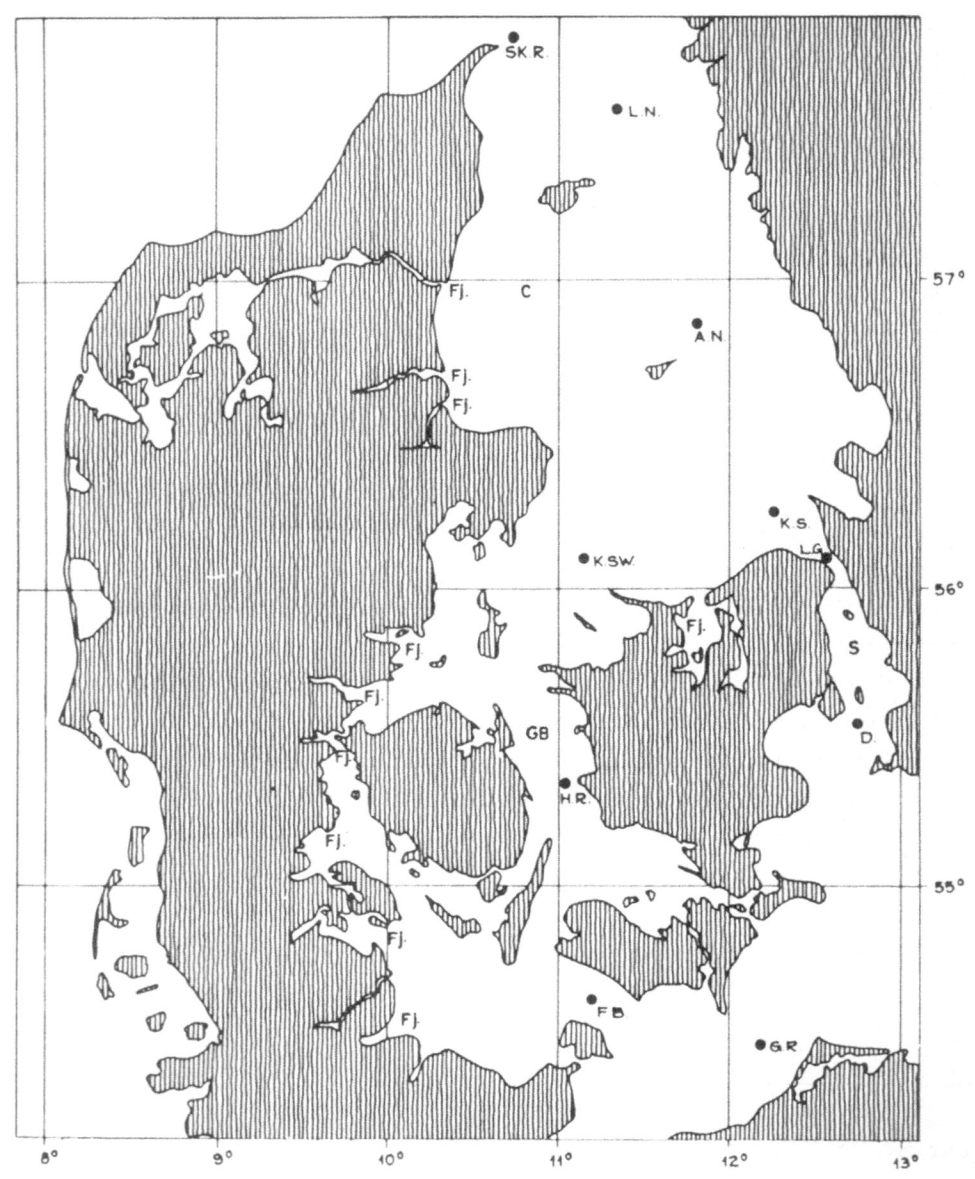

SK R - SKAGENS REV   L G - LAPPEGRUND
L N - LÆSØ NORD
A N - ANHOLT NORD
K SW - KATTEGAT SYDV
K S - KATTEGAT SYD
H R - HALSKOV REV
F B - FEHMERNBELT
G R - GEDSER REV
D   - DROGDEN

Fig. 11.5   The Danish inland waters.
            GB: The Great Belt, S: The Sound,
            C: The Cattegat, Fj: Fiords.
            ●: Lightvessels.

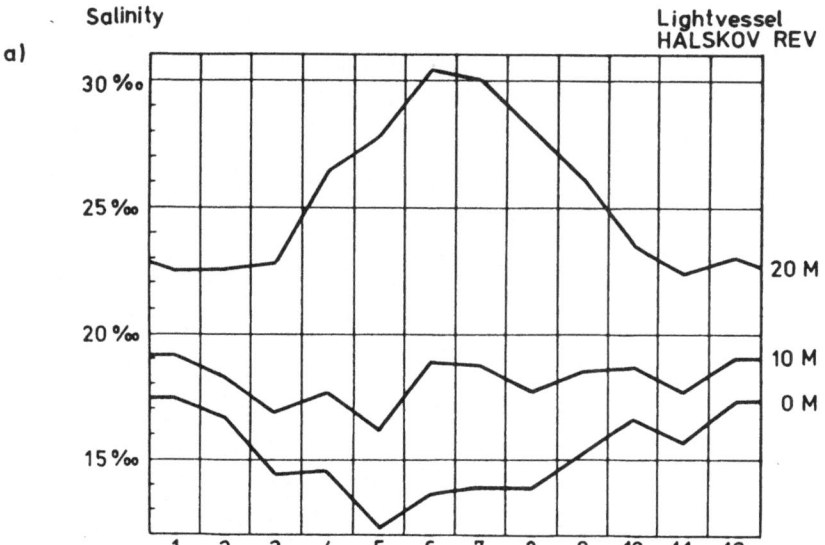

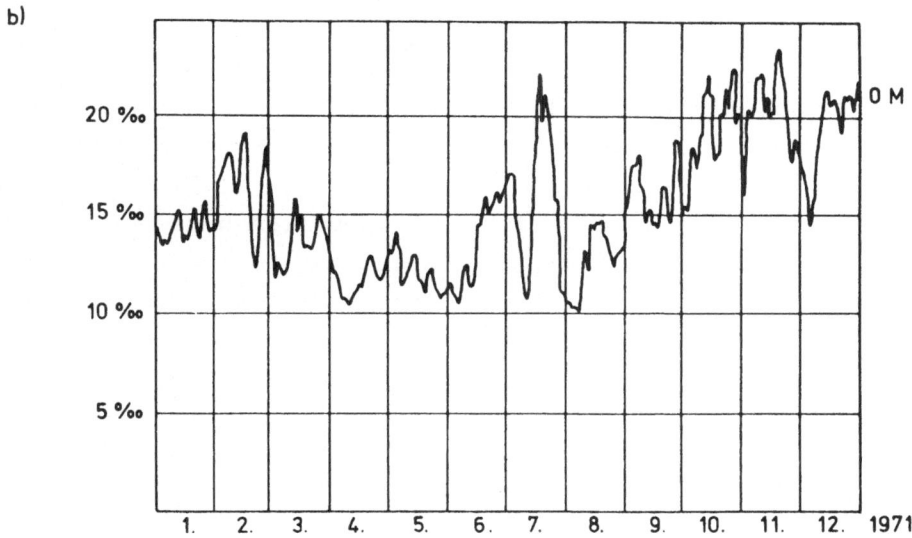

Fig. 11.6 a) The yearly variation of the salinity in
               the Great Belt (average for 30 years).
           b) One year's variation of the salinity in
               the Great Belt.

The Sound: A typical fiord.

The Cattegat: A salt-water wedge.

The Great Belt: A salt-water wedge in the case of normal
                   flow conditions and a well-mixed estuary
                   in the case of flows, directed towards
                   the Baltic.

The many "fiords": Well-mixed estuaries, and fiords.

Due to a very shallow sill between the Sound and the Baltic (8 m depth, approximately at D in Fig. 11.5) the major in-
and outflows to/from the Baltic take place through the Great
Belt. The yearly variation of the salinity here is shown in Fig.
11.6 a (30 years' average) and b (for the year 1971) (from the
Belt project [1976], Nielsen) based on measurements from the
lightvessel H R (Halskov Rev). The general picture is a relative
intense in- and outflow during the winter period, creating more
well-mixed conditions than during the meteorologically more
stable summer period, when a pronounced two-layered system is
more likely. The very rapid changes in the salinity (see Fig.
11.6 b) is primarily caused by a convective transport associated
with a change in the flow-direction, see Fig. 11.7, which is an
extract of the discharge-head loss time series based on the preliminary calculations of the discharge (the Belt project [1976],
Jacobsen) and waterlevel data from the Danish Meteorological
Office. If we try to relate the head loss ΔH from the Cattegat
to the Baltic to the discharge Q by the common specific resistance, K, i.e.

$$\Delta H = K Q^2 \tag{11.1}$$

we obtain the scattered data shown in Fig. 11.8, although we
have omitted all the relatively small values (from Bo Pedersen
[1978 b]). The great variation in the specific resistance is
due to a variation in the level of the interface, see Fig. 7.3,
combined with the great difference in the friction factor for
the flow over the interfacial area (see example 7.1) and for
the flow over the fixed bed. The resistance coefficient is a
minimum for a depth y of about 15 to 20 m (which is the most

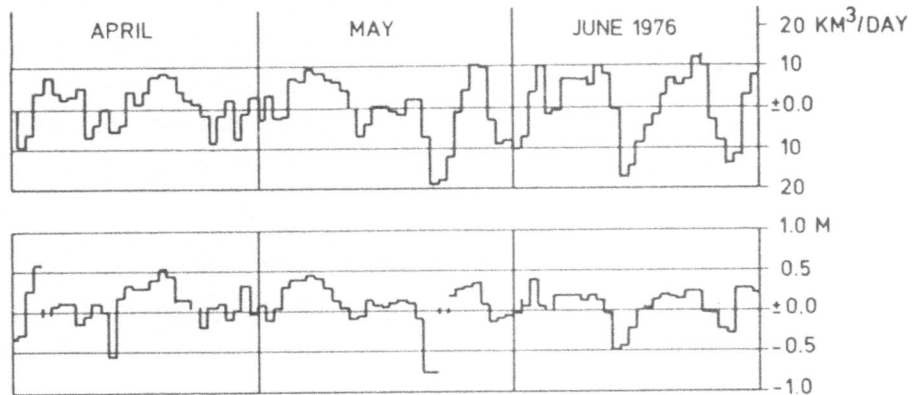

*Fig. 11.7   An extract of the discharge-head loss time
series. From Jacobsen "Bæltprojektet" [1976]
and the Danish Meteorological Office.*

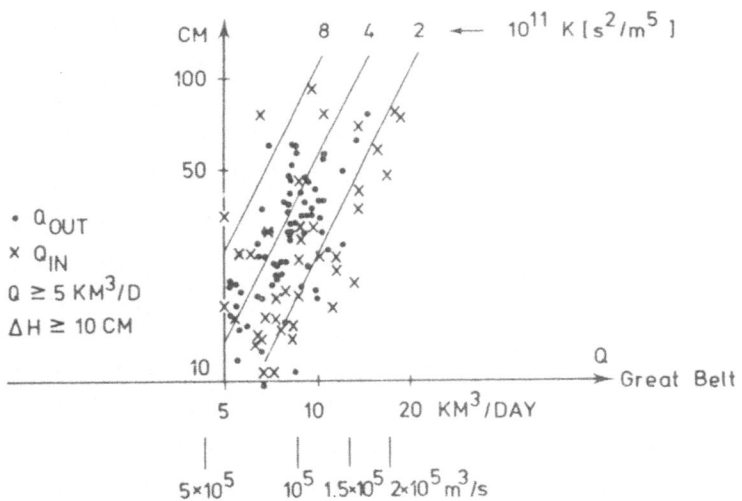

*Fig. 11.8   Dischage Q - Head loss ΔH relation for the
Great Belt. From Bo Pedersen [1978 b].*

135

likely depth) and increases for y decreasing (increase of the velocity) as well as for y increasing (increase of the fixed bed influence), as demonstrated in Bo Pedersen [1978 b].

From Fig. 11.7 we can conclude that typical in- and outflows, respectively, are of the order or magnitude of

$$o(Q_{IN}) \sim o(Q_{OUT}) \sim 10 [KM^3/DAY] \sim 10^5 [m^3/s] \qquad (11.2)$$

which is an order of magnitude higher than the freshwater discharge

$$o(Q_F) \sim 10^4 [m^3/s] \qquad (11.3)$$

from the Baltic.

From the above we may conclude that

i)   making long-term volume budgets for the Baltic based on the calculated (measured velocities integrated) discharges in the Great Belt (and the Sound) is hazardous.

ii) If we want to check our entrainment hypothesis and interfacial shear stress formula, it can only be done by incorporating them in a sophisticated numerical model.

Other types of overflow

The growing concern with the environment has intensified research on three-dimensional surface buoyant jets in the coastal zone and in rivers and lakes. The surface jets may either stem from power stations (cooling water discharge) or sewage outlets. Although theoretical models are numerous in the literature, see for instance one example in Part III, Engelund and Bo Pedersen [1973] the flow field in the recipient water and in the jet itself is often so complicated that it is necessary to rely on numerical modelling, see for instance DHI [1980]. One of the basic requirements for these models is a reliable entrainment function. As the densimetric Froude number is in the intermediate range (from sub- to weak supercritical) where a theoretically determined entrainment function is extremely difficult to outline, it is necessary to rely on laboratory experiments.

As our theory predicts the entrainment velocity (V_E/V) as a function of the friction velocity squared, we have used the densimetric Froude number squared, $\mathbb{F}_{\Delta,F}^2$, based on the friction velocity. This is an important point, as the ordinary densimetric Froude number squared may vary by a factor of ten for the same $\mathbb{F}_{\Delta,F}^2$. With

$$\mathbb{F}_{\Delta,F}^2 = \frac{U_F^2}{\Delta gy} \tag{11.4}$$

as parameter we have shown that all the reported directly measured entrainment data form a single curve, except - of course - for the transition known to take place for a fixed ordinary densimetric Froude number, see Fig. 11.11.

Interflows

The tapping of water from stratified reservoirs (see Fig. 11.9) creates interflows, only a certain stratum being withdrawn. Since the velocity u_0 at the outlet is of the order of magnitude of at least

$$o(u_0) \sim 1 [m/s] \tag{11.5}$$

and since the stratification in the reservoir is for example

$$o(N^2) \sim 10^{-3} [s^{-2}] \tag{11.6}$$

the pressure drop at the outlet

$$o\left(\frac{u_0^2}{2g}\right) \sim 0.1 \ [m] \tag{11.7}$$

cannot normally be counteracted by a change in the internal dynamic pressure

$$o\left(\frac{1}{g} N^2 \frac{y^2}{6}\right) \simeq o\left(\frac{1}{\rho} \frac{\partial \rho}{\partial z} \frac{y^2}{6}\right) \sim o(10^{-5}y^2) \tag{11.8}$$

because that would require a depth of the withdrawn layer of more than a hundred meters. Therefore, in order to obtain a dynamic balance in the system, it is necessary to take the surface

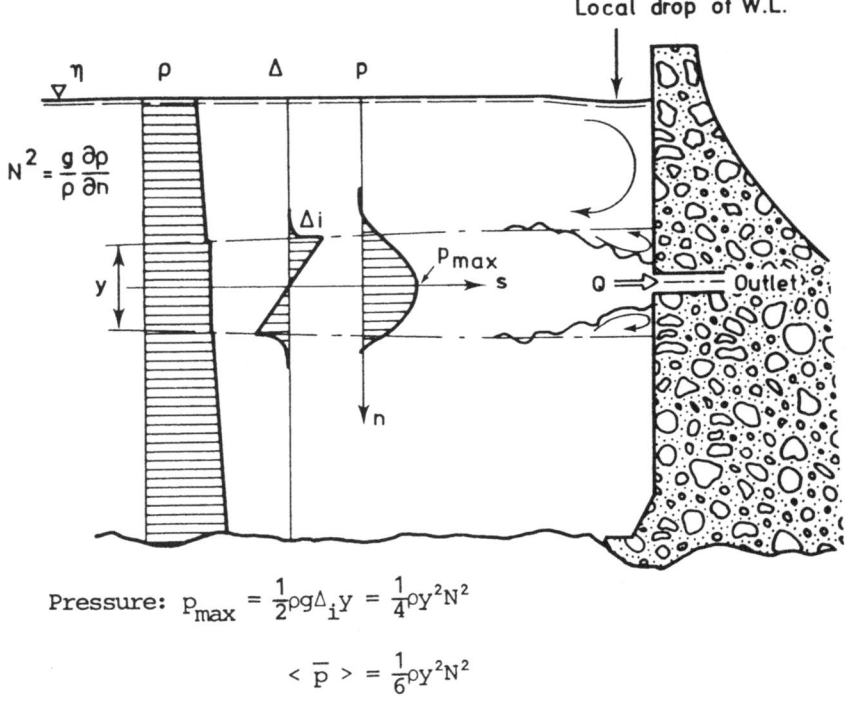

$$\text{Pressure: } p_{max} = \frac{1}{2}\rho g \Delta_i y = \frac{1}{4}\rho y^2 N^2$$

$$< \bar{p} > = \frac{1}{6}\rho y^2 N^2$$

Fig. 11.9 Selective withdrawal from linearly stratified reservoir. Note the drop in water level at the dam. N is the Bruunt-Vaiasäla frequency.

water level drop at the dam into consideration.

As the outlet area, due to the small height and width of the outlet, normally is insignificant compared with the cross-section of the selectively withdrawn layer, the well-mixed layer extends the whole way to the dam, except in the vicinity of the outlet, where a highly intermittent interface may be observed. Accordingly, the flow in the interior is very weak, which - combined with the limited length of the reservoir - means that entrainment into the withdrawn layer is negligible.

Summary on horizontal buoyant flows

Generally, nearly horizontal buoyant flows in nature give the impression that they are highly non-stationary. This behaviour has two explanations. Firstly, due to the very low interfacial shear stress, the energy gradient is very low, too (see example 3.2.1), which implies that the system becomes very sensitive to a change in the external forces. Secondly, the spectrum of external forces which may influence the overflows is extraordinarily wide compared with other flow types. Let us mention some of the most important changes in the boundary conditions:

i) Change of the freshwater (or brackish water) supply. In case of brackish water supply at the boundary, the discharge may even be negative.

ii) Tidal forces are imposed on the whole water column (in fiords with a sill only in the layer above the sill level). This implies that the tidally generated flow is in the ambient fluid as well as in the overflow, see Part III, Møller and Bo Pedersen [1983].

iii) Changes in the wind field over the actual area, which may have a great effect on the flow and, especially, on the mixing processes in overflows with a weak common velocity (e.g. less than 0.1 m/s). A change in the wind field is associated with

iv) a change in the barometric pressure gradient, which may be of importance to overflows of considerable geographic extension, where the Coriolis-effect may become important, too.

v) Finally, the temperature may play an important role to the possible ice capping of the flow, which transfers the overflow to a light roof current (Ch. 8).

If we compare the non-stationary overflow with the other non-stationary buoyancy flows, the most striking difference is to be found in the ambient fluid, which in the case of an overflow normally participates in the non-stationary movements, whereas this is normally not the case for the other buoyancy flows.

Since this is an important point we may elaborate a little the difference in the physical properties of a quasi-stationary flow in which the ambient fluid is at rest and of a quasi-stationary flow in which the ambient fluid is in motion, see

Fig. 11.10. (The term quasi-stationary indicates that the local rate of acceleration $1/g \; \partial V/\partial t$ is small as compared with i.e. the pressure gradient). The mixing process in the light roof current illustrated in Fig. 11.10a is consistent with the theory outlined here, i.e. we have a one-way transport process, namely entrainment (Ch. 6). As the flow is assumed to be quasi-stationary we may relate the entrainment to the instantaneous mean velocity or, if we want to treat the mean flow over a period T, we can use as reference velocity

$$\tilde{V} = (\overline{|V|^3})^{1/3}$$

where $|V|$ = the numerical velocity (production positive).

The reason for us to use the average of the velocity to the third is that the entrainment velocity is proportional to the production of turbulent kinetic energy, which for the subcritical flow treated here is (per unit area)

$$PROD \simeq \tau_i (V - u_i) \sim V^3 \qquad (11.9)$$

For the nearly harmonic varying flow illustrated in Fig. 11.10a

$$\tilde{V} = (2.5 \; V^3)^{1/3} = 1.36 \; V \qquad (11.10)$$

In Fig. 11.10 we have shown the production as a function of time, from which it is clear that the entrainment process is significant for only about a quarter of the period. Therefore, the flow may be treated only as the average over a period T, provided that the length scale for the flow considered is large compared with the progression of a fluid particle during a period T or, expressed in time dimensions, provided that the retention time in the flow section considered is large compared with the period T.

Contrary to the above-mentioned examples, the turbulent production of the non-steady salt water wedge illustrated in Fig. 11.10b is not confined to the upper layer, but is present in the ambient fluid as well. This gives a two-way entrainment,

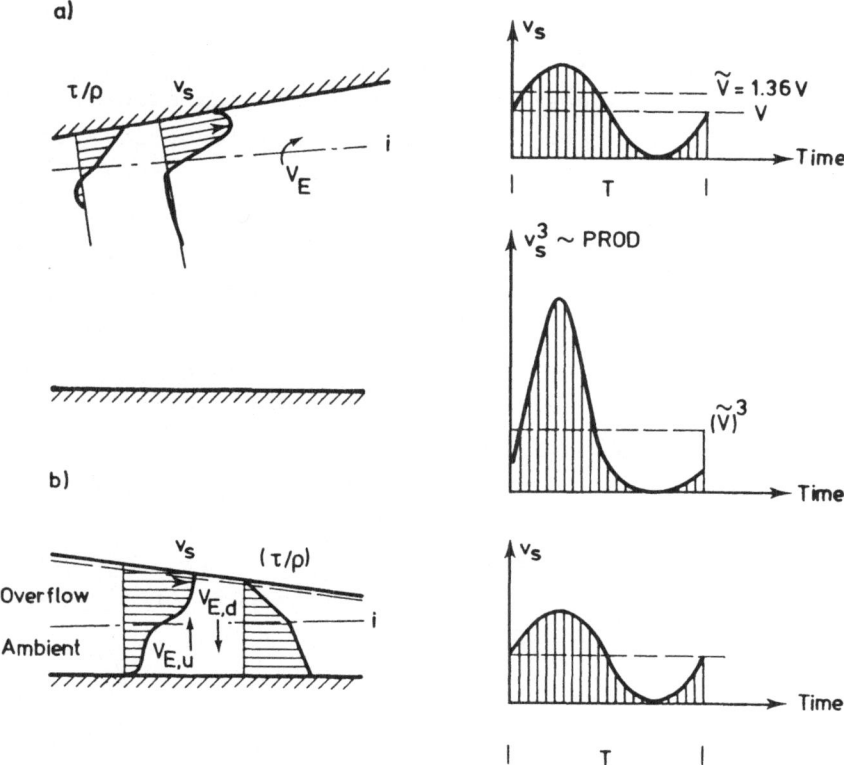

Fig. 11.10 Non-stationary flows in
a) a light roof current, ambient fluid at
rest. Forcing function: the buoyancy
supply.
b) a salt water wedge, ambient fluid in
motion. Forcing function: the tide.

the strength of which is dependent on the ratio of the production in the upper to that in the lower layer, respectively.
Furthermore, the pressure gradient in the ambient fluid is far
from being nil (see Example 3.2.1). The example illustrated
above is extremely simple as compared with the conditions in
the field; therefore, an analytical approach must be almost impossible in the general case. Consequently, we must rely on numerical models as for instance the above mentioned model developed by the Danish Hydraulic Institute (DHI [1977]). Very briefly,
the model deals with a two-dimensional two-layered flow. The
mixing has been treated as a two-way entrainment.

By a number of examples we have now illustrated that it is
nearly impossible to make an analytical approach to the general
case of an overflow. The examples chosen are representative of
the flows in nature, apart from the fact that they have been
treated very crudely - nature is more complicated than that.
Another complication created by the non-stationarity is the
movement of fronts, i.e. regions across which the hydrographical
properties of the water change their character abruptly.

The examples of overflows which we have gone through, have
shown us that we should be more concerned with the basic physical problems than with outlining a mathematical description of
a non-existent well-behaved steady flow.

Therefore, we shall restrict ourselves to treat some of
the basic physical problems, namely the entrainment (Ch. 6) and
the interfacial shear stress (Ch. 7). For reasons explained
above, a theoretical outline of the entrainment function will
be given for subcritical flow only.

Subcritical flow

If we want to investigate the entrainment of a free surface buoyant flow, we must first of all make sure that the turbulent production stems from the interfacial shear stress, i.e.
that the effects from the side walls are negligible. The production per unit area due to an overflow is (Eq. 4.2.14)

$$\text{PROD}_{\text{wide flume}} \simeq \tau_i (V - u_i) \simeq \rho \frac{\tau_i / \rho}{(V - u_i)^2} (V - u_i)^3 \quad (11.11)$$

If we take a flume with a small width to depth ratio
(which is the most common one) the production is:

$$\text{PROD}_{\text{narrow flume}} \simeq \tau V \simeq \rho \frac{\tau / \rho}{V^2} V^3 \quad (11.12)$$

which is an order of magnitude higher. Therefore, if we intend
to obtain pure overflow entrainment data, the depth of the upper layer must be extremely small compared with the width.

For a horizontal subcritical overflow the entrainment is
determined by the bulk flux Richardson number to be (Ch. 6)

$$\frac{V_E}{V} = 2\left[\frac{f_i}{2}\left(\frac{U_m}{V} - \frac{u_i}{V}\right)^2 \left(1 - \frac{u_i}{V}\right)\right] \mathbb{R}_f^T \ \mathbb{F}_\Delta^2 \tag{11.13}$$

where the densimetric Froude number squared here is defined as

$$\mathbb{F}_\Delta^2 = \frac{V^2}{\xi \Delta gy} \tag{11.14}$$

If we introduce the friction velocity U_F defined by

$$U_F = \sqrt{\tau_i/\rho} = \sqrt{f_i/2(U_m - u_i)^2} \tag{11.15}$$

and accordingly, a densimetric Froude number squared, based on this friction velocity as

$$\mathbb{F}_{\Delta,F}^2 = \frac{U_F^2}{\Delta gy} \tag{11.16}$$

the entrainment function may be rewritten to yield

$$\frac{V_E}{V} = \frac{1}{\xi} \ 2 \ \mathbb{R}_f^T \left(1 - \frac{u_i}{V}\right) \mathbb{F}_{\Delta,F}^2 \tag{11.17}$$

Further, if we introduce the experimentally verified value $u_i/V = 0.56$ and the flux Richardson number $\mathbb{R}_f^T = 0.045$ (Ch.5) we obtain

$$\frac{V_E}{V} = \frac{1}{\xi} \ 0.040 \ \mathbb{F}_{\Delta,F}^2 \tag{11.18}$$

where ξ is equal to one in fully developed flow only (long flumes and in the field). In this theory the entrainment function - non-dimensionalized by the mean velocity - is proportional to the friction velocity squared, not to the mean velocity squared as is common in the literature. This is an important finding, because the ratio of the friction velocity to the mean velocity varies significantly from a low Reynolds number model test to a high Reynolds number field test, see Ch. 7. This difference is elucidated in the entrainment function drawn in Fig. 11.11. Besides the field measurements by Fukushima et al. [1969], we have the numerical model by DHI [1977] to verify the theory.

143

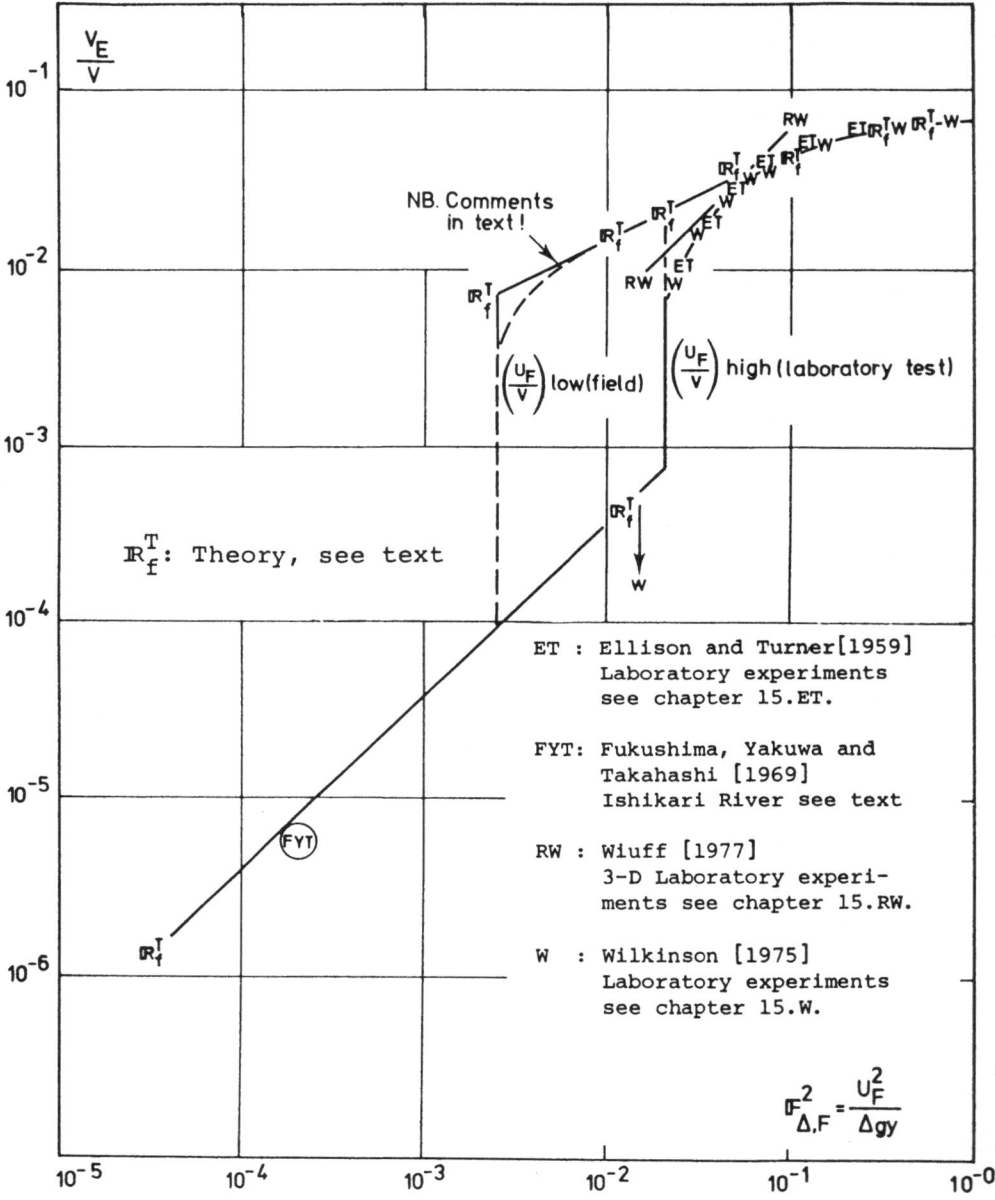

Fig. 11.11   The entrainment function for free surface
buoyant flow.

Supercritical flow

The entrainment into supercritical buoyancy flow is significant and of great practical importance. Recently, Sehested, DHI [1982] has performed a detailed analysis of entrainment into horizontal buoyant jets, based on the $\mathbb{R}_f^T$ - hypothesis, and up-to-date measurements.

It seems appropriate to end this chapter with a

WARNING

We must emphasize that an uncritical application of the above outlined theory to a highly non-stationary flow may lead to considerable errors, because

    i)   the friction factor in highly non-stationary flows in the field is normally an order of magnitude higher than the friction factor for a stationary flow (growing boundary layer).

   ii)   the centre of gravity movement of the entrained fluid in a rapidly varying flow (in time or space) is normally an order of magnitude lower than in a gradually varying, stationary flow.

  iii)   The pressure gradients (and thus the shear stresses) and the velocities are normally significant in the ambient fluid, producing a non-negligible turbulence and thus a downward entrainment.

Conclusion concerning rapidly varying buoyancy flows in relation to the theory outlined: If the changed friction velocity and the changed centre of gravity movement of the entrained fluid are not taken into account, errors of several orders of magnitude are likely to appear in the results.

12. VERTICAL BUOYANT JETS AND PLUMES

A vertical buoyant jet is the flow created by a source of fluxes of mass, momentum, and buoyancy flowing into an ambient fluid. The jet is primarily driven by the momentum flux, while the plume is driven by buoyancy only.

As jets are associated with a high momentum flux, created by converting a high pressure into kinetic energy through a small opening, examples from the geophysical field of vertical buoyant jets are rare, while they are extremely common as a product of activity of man.

From Iceland and other volcanic active regions vertical buoyant jets of steam and boiling water (Geysers) are a well-known scenery in nature, and even jets of melted lava and lapilli occur from time to time.

From the civil engineering field a multiport diffusor is shown in Fig. 12.1, in which jets from numerous ports (equally spaced along the top of the outlet pipe) merge and form a two-dimensional buoyant jet. Such a system can be found in shore-sited power plants where coastal water is used for cooling purposes and in sewage plants where the coastal water is used as recipient. The warm water (excess of heat) and the sewage water (excess of pollution) is discharged through a diffusor in order to achieve a certain dilution before disposal. The degree of dilution is determined either by political restrictions (environmental protection) or/and by economic considerations concerning the efficiency of the cooler or the diffusor. The physical factor determining the degree of dilution is the mixing of the buoyant jet with the ambient water and, consequently, the input of energy into the system. To optimize the diffusor, it is therefore necessary to know the efficiency of the mixing process, i.e. the ratio of the energy gained (the dilution) to the energy input.

A further example of great practical importance is the smoke issued from a chimney which creates a vertical buoyant circular jet-plume flow. This flow is normally superimposed by a log-linear wind velocity profile in a stratified atmosphere, see Fig. 12.2. Due to the stable stratification, or perhaps com-

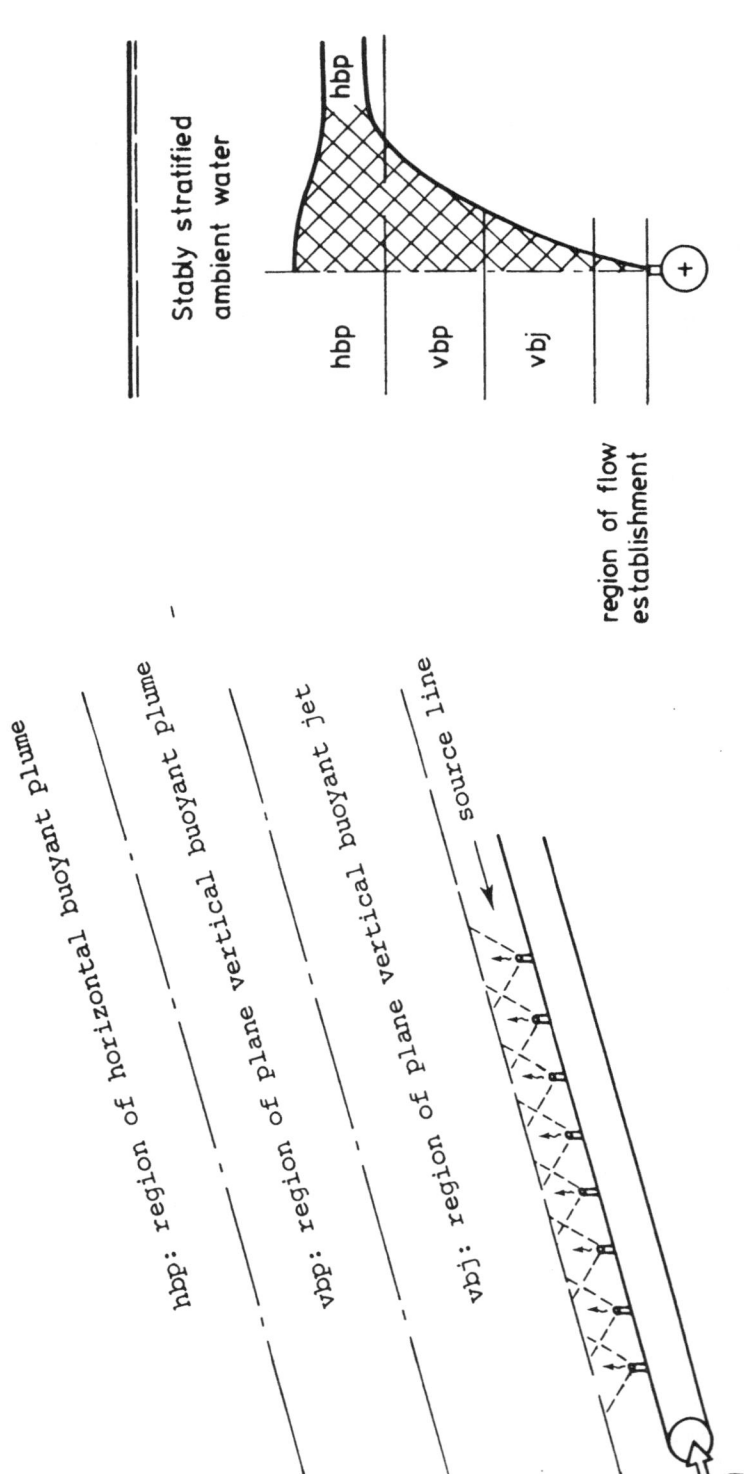

Fig. 12.1 Sketch of the different zones in a common diffusor situated in a stable stratified ambient fluid.

bined with an inversion capping the atmospheric boundary layer,
such a chimney can cause heavy air pollution. The dilution of
the smoke - and, accordingly, the degree of pollution concentration - is again a combined political/economic (and of course a
meteorological) question.

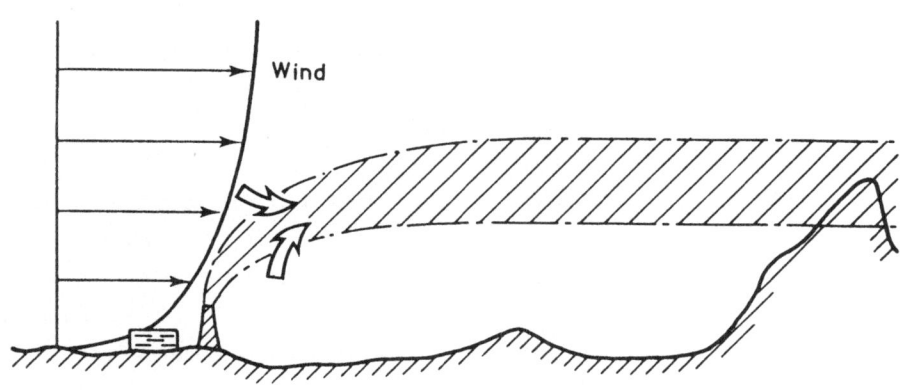

*Fig. 12.2 Air pollution produced by the smoke from a
chimney in a stably stratified ambient atmosphere with a log-linear velocity profile.*

Vertical buoyant jets are dominated by the momentum flux
at the beginning. Due to the entrainment the densimetric Froude
number decreases steadily until the final steady state is reached, which is a vertical buoyant plume with constant densimetric
Froude number, driven by the buoyancy forces. As the densimetric Froude number for vertical buoyant jets is in the supercritical flow range

$$\mathbb{F}_{\Delta, \text{plume}} < \mathbb{F}_{\Delta, \text{v b jet}} < \infty \qquad (12.1)$$

the entrainment process is associated with vortex formation
(compare the description given in Ch. 6). As the entrainment is
horizontal, the local gain in potential energy is nil, which
means that the bulk flux Richardson number, $\mathbb{R}_f^T$ (Ch. 5), in this
case is reduced to the ratio of the gain in turbulent kinetic
energy due to entrainment to the production corrected for the
transfer of turbulent kinetic energy, which simply ends up in

a constant relative entrainment velocity

$$\frac{V_E}{V} = 0.5 \ \mathbb{R}_f^T = 0.09 \tag{12.2}$$

applicable for jets as well as for plumes.

Comments:

This statement is in contradiction with the "established practice", but has been verified in Bo Pedersen [1980], and in Henriksen, Haar and Bo Pedersen [1982] - see Part III. Other literature will often present an entrainment velocity for plumes which is double the value given here. The explanation is that plumes are subject to meandering (see Fig. 6.2 c), which is erroneously interpreted as being due to entrainment - and hence the entrainment is overestimated (by a factor of two!). Similarly, the Eulerian measured velocities and reduced mass are under-estimated - again by a factor of two. We shall not go further into that matter here, but point out that we are dealing with Lagrangian (momentaneous) properties of the jets/plumes in the present chapter.

————

Plumes are characterized by a constant densimetric Froude number $\mathbb{F}_{\Delta, plume}$ which indicates a constant ratio of inertia forces to buoyancy forces. This state of equilibrium is connected with the balance between the driving gravitational force (buoyancy) and the retarding forces (friction and acceleration of the entrained fluid). Therefore, similarity may be expected to exist for the velocity and the density profiles, respectively.

When a dense bottom current is tilted to a vertical position it becomes half a plume, and hence the densimetric Froude number for plumes constitutes the *upper limit* of dense bottom currents and light roof currents and the *lower limit* of vertical jets, as illustrated in Fig. 12.3, i.e.

$$0 < \mathbb{F}_{\Delta, \text{dense bottom currents}} \lesseqgtr \mathbb{F}_{\Delta, plume}$$

$$= \text{const} \lesseqgtr \mathbb{F}_{\Delta, \text{vertical buoyant jets}} < \infty$$

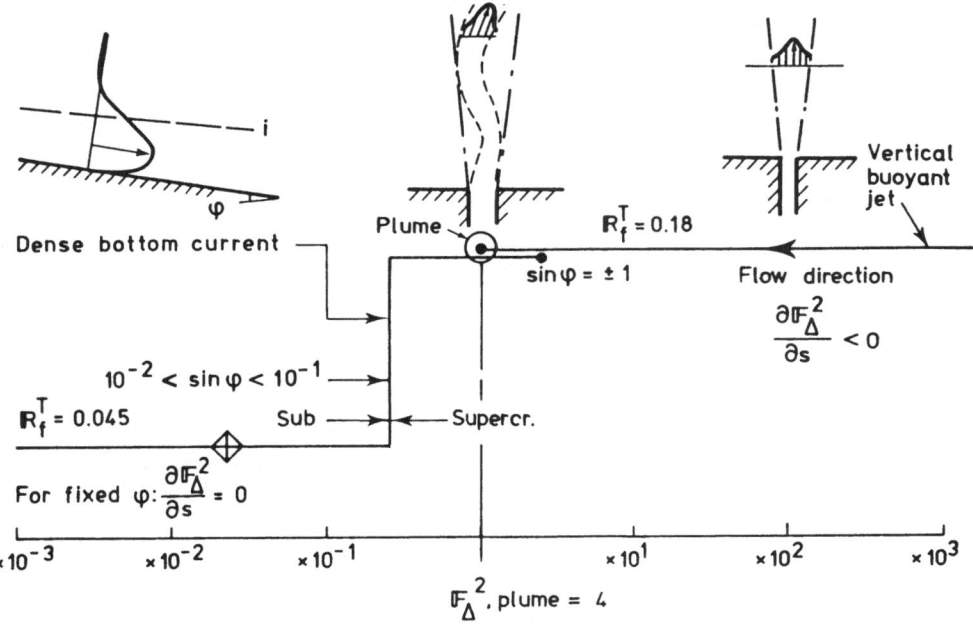

Fig. 12.3 *A demonstration of the central position of plumes being the final stage for i) all vertical buoyant jets and ii) for dense bottom currents (or light roof currents), when the bottom becomes vertical.*

In example 4.2.2, the momentum equation for a vertical jet was outlined.

As plumes constitute the final stage in a vertical buoyant jet, we can directly apply the findings in Example 4.2.2 and write the equation of motion for a plume, taking the pertubations from the meandering as having a small effect on the flow-directed balance

$$\frac{\partial}{\partial s}\,(\rho \alpha' V q) \;=\; \Delta \rho g y \tag{12.3}$$

This momentum equation shows that we have a steady increase in momentum created by the buoyancy excess.

The constant densimetric Froude number for plumes implies that the velocity is constant (see exercise 4.1.1), and hence equation 12.3 may be rewritten as

$$\frac{\partial y}{\partial s} = \frac{\Delta g y}{\alpha' V^2} \qquad (12.4)$$

where the continuity equation can be used to eliminate $\partial y / \partial s$

$$2 V_E = \frac{\partial q}{\partial s} = V \frac{\partial y}{\partial s} \qquad (12.5)$$

Therefore we find the constant densimetric Froude number squared for plumes as

$$\mathbb{F}_{\Delta}^2, \text{plume} = \frac{V}{\alpha' 2 V_E} = 4.0 \qquad (12.6)$$

Notice that the spreading for jets was (Example 4.2.2)

$$\left(\frac{\partial y}{\partial s}\right)_{\text{jets}} = 4 \frac{V_E}{V} = 0.36 \qquad (12.7)$$

while the spreading (the Lagrangian) for plumes is

$$\left(\frac{\partial y}{\partial s}\right)_{\text{plumes}} = 2 \frac{V_E}{V} = 0.18 \qquad (12.8)$$

(where the Eulerian spurious spreading is - by coincidence - 0.36, as for jets, where no meandering takes place).

In Fig. 12.4 an experiment is shown which creates a plume on the basis of two vertical dense bottom currents. From Ch. 8 we know that the Richardson number for dense bottom current depends on the total friction factor. A smooth transition is obtained by letting the wall friction be negligible.

Let us take the interfacial friction factor (related to the mean velocity V) in our experiment to be

$$(f/2) \sim 2 \times 10^{-2} \qquad (12.9)$$

which yields

$$\mathbb{Ri}_{d\ b\ c} = \frac{\frac{1}{2} \Delta g q}{V^3} = 0.12 \qquad (12.10)$$

and, accordingly, the plume Richardson number

$$\mathbb{Ri}_{pl} = 2 \mathbb{Ri}_{d\ b\ c} = 0.24 = \frac{1}{\mathbb{F}_{\Delta, pl}^2} \qquad (12.11)$$

in good agreement with our calculation above. Ensuring that the initial plume is laterally stable, this armchair experiment makes our meandering interpretation of the measurements probable (the Eulerian measurements yield $\mathbb{R}i, pl = 1.4$, i.e. a factor 6 higher than the Lagrangian value!)

Hitherto, we have been concerned with two-dimensional jets/plumes only. In part III of the present lecture notes an example of a circular jet/plume is given, Henriksen, Haar and Bo Pedersen [1982]. Although the flow treated is rather special - the mass excess is due to suspended particles - the basic equations outlined apply to ordinary jets/plumes, too. In practical waste

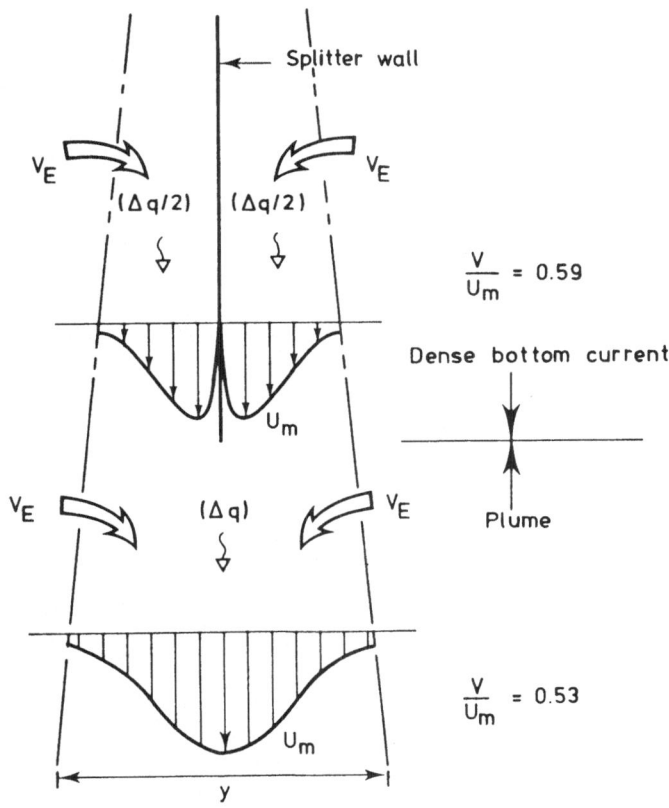

Fig. 12.4   A theoretical experimental set-up which creates a plume by combining two dense bottom currents when they issue from the end of the infinitesimally thick splitter wall.

water discharge design, one is most concerned with the dilution of the harmful effluent, and to that end we may refer to Fig. 3 in the above mentioned paper, where a numerical solution to the horizontally discharged jet is shown.

It is rather amazing that the outlet Froude number - which is proportional to the kinetic energy input into the diffusor - only has a very weak influence on the dilution. Within a wide range, we may simply - as a "rule of thumb" - estimate the cross-average dilution S as

$$\frac{1}{3} \frac{y}{r_0} \leq S < \frac{1}{2} \frac{y}{r_0} \qquad (12.12)$$

where r_0 is the diffusor radius and y the vertical movement of the jet. The cross-average dilution is 1.5 times the centerline dilution.

APPENDIX - THE STATE OF SEAWATER

Water - as it may be found in the ocean, in estuaries and in lakes - has a density ρ [kg/m³] which is determined by the temperature T[°C], the salinity S[°/oo] (to be defined below), the pressure p [bar = 10^5 N/m²] and in some cases by the content of suspended particles.

The absolute salinity S_A is defined as

$$S_A = \frac{\text{mass of dissolved salt}}{\text{mass of seawater}} \tag{A1}$$

In the Atlantic Ocean - for example - approximately 35 kg of salt are dissolved per 1000 kg of sea-water, while in the Baltic Sea the salt content is only 8 to 10 kg per 1000 kg seawater, due to the dilution by river runoff.

As S_A cannot be measured directly (only by titration of the sea-water sample), a new definition of the salinity has been given by Unesco [1981a]. The background for the alternative definition is that the ion-composition of the different sort of salts dissolved is insignificantly varying in the ocean, which means that the conductivity is an excellent - and practical - measure for the salinity. Hence, the socalled practical salinity S is defined by a UNESCO Committee [1981a] on the basis of the conductivity of the sea-water sample, see the following page (which is a reprint from the report).

The next step is to relate the density ρ to the temperature T and the salinity S, which can only be done imperically based on a huge number of accurate measurements. This great task has been accompliced by UNESCO, too, [1981b]. The results are reprinted on the following pages (one for low pressure, another for high pressure conditions, respectively).

THE PRACTICAL SALINITY, 1978

DEFINITION

The practical salinity, symbol S, of a sample of
seawater, is defined in terms of the ratio K_{15} of the electrical
conductivity of the seawater sample at the temperature of 15°C
and the pressure of one standard atmosphere, to that of a potassium
chloride (KCl) solution, in which the mass fraction of KCl is
32.4356×10^{-3}, at the same temperature and pressure. The K_{15} value
exactly equal to 1 corresponds, by definition, to a practical
salinity exactly equal to 35. The practical salinity is defined in
terms of the ratio K_{15} by the following equation

$$S = 0.0080 - 0.1692 \, K_{15}^{1/2} + 25.3851 \, K_{15}$$

$$+ 14.0941 \, K_{15}^{3/2} - 7.0261 \, K_{15}^{2} + 2.7081 \, K_{15}^{5/2}$$

formulated and adopted by the Unesco/ICES/SCOR/IAPSO Joint Panel
on Oceanographic Tables and Standards, Sidney, B.C., Canada,
1 to 5 September 1980 and endorsed by the International Association
for the Physical Sciences of the Ocean (IAPSO) in December 1979,
the International Council for the Exploration of the Sea (ICES) in
October 1979, the Scientific Committee on Oceanic Research (SCOR)
in September 1980 and the Intergovernmental Oceanographic Commission
(IOC) of Unesco in June 1981. This equation is valid for a
practical salinity S from 2 to 42.

Reprint, Unesco [1981a]

The One Atmosphere International Equation
of State of Seawater, 1980

Definition

The density (ρ, kg m^{-3}) of seawater at one standard atmosphere (p = o) is to be computed from the practical salinity (S) and the temperature (t, °C) with the following equation :

$$\rho(S,t,o) = \rho_w + (8.244\ 93 \times 10^{-1} - 4.0899 \times 10^{-3}\ t$$

$$+ 7.6438 \times 10^{-5}\ t^2 - 8.2467 \times 10^{-7}\ t^3 + 5.3875 \times 10^{-9}\ t^4)S$$

$$+(-5.724\ 66 \times 10^{-3} + 1.0227 \times 10^{-4}\ t - 1.6546 \times 10^{-6}\ t^2)S^{3/2}$$

$$+ 4.8314 \times 10^{-4}\ S^2$$

where ρ_w, the density of the Standard Mean Ocean Water (SMOW) taken as pure water reference, is given by·

$$\rho_w = 999.842\ 594 + 6.793\ 952 \times 10^{-2}\ t - 9.095\ 290 \times 10^{-3}\ t^2$$

$$+ 1.001\ 685 \times 10^{-4}\ t^3 - 1.120\ 083 \times 10^{-6}\ t^4$$

$$+ 6.536\ 332 \times 10^{-9}\ t^5$$

The one atmosphere International Equation of State of Seawater, 1980 is valid for practical salinity from 0 to 42 and temperature from -2 to 40°C.

Reprint, Unesco [1981b]

The High Pressure International Equation
of State of Seawater, 1980

Definition

The density (ρ, kg m^{-3}) of seawater at high pressure is to be computed from the practical salinity (S), the temperature (t, °C) and the applied pressure (p, bars) with the following equation :

$$\rho(S,t,p) = \frac{\rho(S,t,o)}{1 - p/K(S,t,p)}$$

where $\rho(S,t,o)$ is the one atmosphere International Equation of State 1980, given on the preceding front page and $K(S,t,p)$ is the secant bulk modulus given by

$$K(S,t,p) = K(S,t,o) + Ap + Bp^2$$

where

$$K(S,t,o) = K_w + (54.6746 - 0.603\ 459\ t + 1.099\ 87 \times 10^{-2}\ t^2$$

$$- 6.1670 \times 10^{-5}\ t^3)S + (7.944 \times 10^{-2} + 1.6483 \times 10^{-2}\ t$$

$$- 5.3009 \times 10^{-4}\ t^2)\ S^{3/2}\ ,$$

$$A = A_w + (2.2838 \times 10^{-3} - 1.0981 \times 10^{-5}\ t - 1.6078 \times 10^{-6}\ t^2)S$$

$$+ 1.910\ 75 \times 10^{-4}\ S^{3/2}$$

$$B = B_w + (-9.9348 \times 10^{-7} + 2.0816 \times 10^{-8}\ t + 9.1697 \times 10^{-10}\ t^2)S$$

the pure water terms K_w, A_w and B_w of the secant bulk modulus are given by

$$K_w = 19\ 652.21 + 148.4206\ t - 2.327\ 105\ t^2 + 1.360\ 477 \times 10^{-2}\ t^3$$

$$- 5.155\ 288 \times 10^{-5}\ t^4$$

$$A_w = 3.239\ 908 + 1.437\ 13 \times 10^{-3}\ t + 1.160\ 92 \times 10^{-4}\ t^2$$

$$- 5.779\ 05 \times 10^{-7}\ t^3$$

$$B_w = 8.509\ 35 \times 10^{-5} - 6.122\ 93 \times 10^{-6}\ t + 5.2787 \times 10^{-8}\ t^2$$

The high pressure International Equation of State of Seawater, 1980 is valid for practical salinity from 0 to 42, temperature from -2 to 40°C and applied pressure from 0 to 1000 bars.

Reprint, Unesco [1981b]

As the equation of state for sea-water is rather complicated, the ρ-variation is shown in a diagram, Fig. A1. Within a large range of salinity variations it is a good approximation to relate the density linearly to the salinity

$$\rho = \rho(S_0, T_0) \left[1 + \beta(S - S_0)\right] \qquad (A2)$$

where

$$\beta = \frac{1}{\rho} \frac{\partial \rho}{\partial S} \simeq 0.8 \ [\text{per } ^o/oo] \qquad (A3)$$

is a "coefficient of expansion".

The similar linear approximation for the temperature dependency

$$\rho = \rho(S_0, T_0) \left[1 + \alpha(T - T_0)\right] \qquad (A4)$$

where the coefficient of expansion is

$$\alpha = -\frac{1}{\rho} \frac{\partial \rho}{\partial T} \ [\text{per } ^oC] \qquad (A5)$$

is much more dubious (see Fig. A1), especially in the low temperature range.

In density stratified flows the variation in space and time of the density is of crucial importance. Therefore, we may gain in computational accuracy by introducing the deviation of the density from some standard density, say 1000 $[kg/m^3]$. This quantity - denoted sigma -

$$\sigma_{S,T} = (\rho_{S,T} - 1000) \ [kg/m^3] \qquad (A6)$$

is normally within the range of zero to say 30 $[kg/m^3]$.

σ is mostly used within oceanography. In hydraulic engineering it is more common to use the dimensionless reduced density (delta)

$$\Delta = \frac{\rho - \rho_R}{\rho_R} \simeq \frac{\rho - \rho_R}{\rho} \qquad (A7)$$

where ρ_R is a reference density, for instance - but not always - 1000 $[kg/m^3]$.

Provided the linear approximation for the density - salinity relation can be applied, equations (A3) and (A7) yield

$$(\rho - \rho_R) = \Delta\rho_R = \rho_R \beta (S - S_R)$$

or $\qquad \Delta = \beta(S - S_R)$ (A8)

Similarly we obtain

$$\Delta = \alpha(T - T_R)$$ (A9)

where index R stands for reference property. Equations (A8) and (A9) are applied extensively in the lecture notes.

Multiplied by the acceleration of gravity g, the reduced acceleration of gravity appears

$$g' = \Delta g$$ (A10)

which is a very important quantity in density stratified flows.

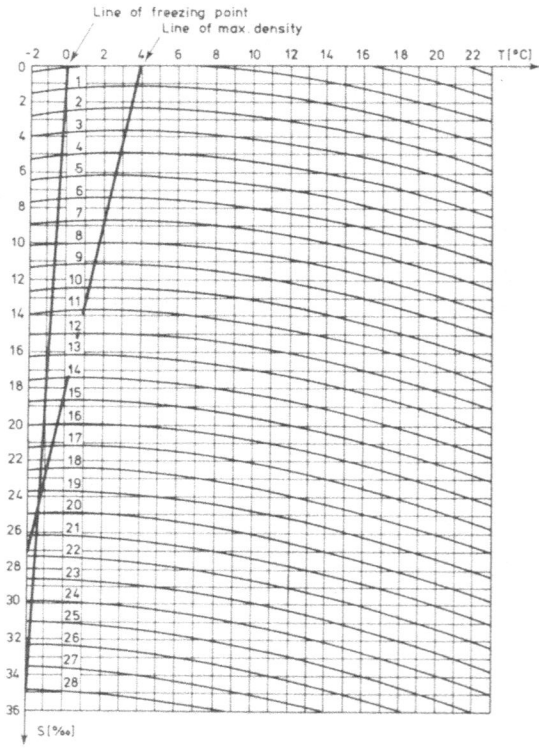

Diagram A1. σ-sea-water for various temperatures and salinities.

REFERENCES

Bengtsson, L. [1978]:
"Wind induced circulation in lakes". Nordic Hydrology,
Vol. 9, pp. 75-94.

Bo Pedersen, Fl. [1972]:
The friction factor for a two-layer stratified flow,
immiscible and miscible fluids". Technical University of
Denmark, Institute of Hydrodynamics and Hydraulic
Engineering, Progress Report No. 27, pp. 3-13.

Bo Pedersen, Fl. [1974a]:
"Interfacial mixing in two-layer stratified flow".
DCAMM report No. 74, Danish Center for applied
Mathematics and Mechanics, Technical University of
Denmark.

Bo Pedersen. Fl. [1974b]:
"Opsamling af olie i vandløb og havne". Lecture Notes [in
Danish]. Technical University of Denmark, Institute of
Hydrodynamics and Hydraulic Engineering.

Bo Pedersen. Fl. [1976]:
"The flux Richardson number and the entrainment
function". Technical University of Denmark, Institute of
Hydrodynamics and Hydraulic Engineering, Progress Report
No. 39, pp. 23-28.

Bo Pedersen, Fl. [1977]:
"On dense bottom currents in the Baltic Deep Water".
Nordic Hydrology, Vol. 8, pp. 297-316.

Bo Pedersen, Fl. [1978a]:
"A brief review of present theories of fjord dynamics".
In: Hydrodynamics of Estuaries.and Fjords. Ed. by
J Nihoul, Elsevier. pp. 407-422.

Bo Pedersen, Fl. [1978b]:
"On the influence of a bridge across the Great Belt on
the hydrography of the Baltic Sea". 11th Conference of
the Baltic Oceanographers, Rostock, DDR. Paper No. 22,
pp.366-377.

Bo Pedersen, Fl. [1980]:
"A monograph on turbulent entrainment and friction in
two-layer stratified flow". Technical University of
Denmark, Institute of Hydrodynamics and Hydraulic
Engineering, Series Paper 25.

Bradbury, L.J.S. [1965]:
"The structure of a self-preserving turbulent plane jet".
Journal of Fluid Mechanics, Vol. 23, pp. 31-64.

Browand, F.K. and Wang, Y.H. [1972]:
"An experiment on the growth of small disturbances at the
interface between two streams of different densities and
velocities". International Symposium on Stratified Flows,
Novosibirsk, Proc. 491-498.

Brown, G.L. and Roshko, A. [1971]:
"The effect of density difference on the turbulent mixing layer". AGARD Conference Proceedings on Turbulent Shear Flows, London, AGARD-CP-93, Session IV, 23-1 - 23-12.

Bæltprojektet [1976]:
"Interim report on the Danish Belt project (in Danish)". Authors quoted: Nielsen, A., Jacobsen, T. Publ. by Miljøstyrelsen, Kampmannsgade 1, DK-1604 Copenhagen, Denmark.

Cross, R. H. and Hoult, D.P. [1971]:
"Collection of oil slicks". ASCE Proceedings, Journ. of the Waterways, Harbors and Coastal Eng. Div. Vol. 97, No. WW2.

Deardorff, J.W., Willis, G.E. and Lilly, D.K. [1969]:
"Laboratory investigation of non-steady penetrative convection". Journal of Fluid Mechanics, Vol. 35, pp. 7-31.

DHI, Report [1977]:
Bæltprojektet. Matematiske modeller af Store Bælt og Øresund. Slutrapport. [In Danish]. Danish Hydraulic Institute, DK-2970 Hørsholm, Denmark.

DHI, Report [1980]:
"On the spreading and mixing of buoyant plumes". Research Report TR 133/312-80.132.

Edward, A. and Edelsten, D.J. [1977]:
"Deep water renewal of Loch Etive: A three basin Scottish fjord". Estuarine and Coastal Marine Science, Vol. 5, pp. 575-595.

Ellison, T.H. and Turner, J.S. [1959]:
"Turbulent entrainment in stratified flows". Journal of Fluid Mechanics, Vol. 6, pp. 423-448.

Engelund, F.A. and Bo Pedersen, Fl. [1982]:
"Hydraulik" (in Danish). Den Private Ingeniørfond, Technical University of Denmark.

Farmer, D.M. [1975]:
"Penetrative convection in the absence of mean shear". Quarterly Journal of the Royal Meteorological Society, Vol. 101, pp. 869-891.

Fukushima, H., Yakuwa, I. and Takahashi, S. [1968]:
"Salinity diffusion at the interface of stratified flow in an estuary". 13th Congress of the International Association for Hydraulic Research, IAHR, Kyoto, Vol. 3, pp. 191-197.

Gade, H.G. [1971]:
"Deep water renewal in a sill fjord". Geofysisk Institutt, University of Bergen, Norway.

Gade, H.G. [1973]:
"Deep water exchanges in a sill fjord: a stochastic
process". Journal of physical Oceanography, Vol. 3,
pp. 213-219.

Georgiev, B.V. [1972]:
"Some experimental investigation on turbulent
characteristics of stratified flows". International
Symposium on Statified Flows, Novosibirsk, Proc,
507-514.

Hamblin, P.F. [1977]:
"Short-period internal waves in the vicinity of a
river-induced shear zone in a fjord lake". Journal of
Geophysical Research, Vol. 82, pp. 3167-3174.

Hoult, D.P. [1972]:
"Oil spreading on the sea". Annual Review of Fluid
Mechanics, Vol. 4, pp. 341-368.

Kantha, L.H., Phillips, O.M. and Azad, R.S. [1977]:
"On turbulent entrainment at a stable density interface".
Journal of Fluid Mechanics, Vol. 79, pp. 753-768.

Kato, H. and Phillips, O.M. [1969]:
"On the penetration of a turbulent layer into stratified
fluid". Journal of Fluid Mechanics, Vol. 37, pp. 643-655.

Kim, H.T., Kline, S.J. and Reynolds, W. [1971]:
"The production of turbulence near a smooth wall in a
turbulent boundary layer". Journal of Fluid Mechanics,
Vol. 50, pp. 133-160.

Kline, S.J., Reynolds, W.C., Schraub, F.A. and Runstadler, P.W.
[1967]:
"Structure of turbulent boundary layers". Journal of
Fluid Mechanics, Vol. 30, pp. 741-773.

Knudsen, M. [1900]:
"Ein hydrographischer Lehrsatz". Ann. Hydr. Mar. Met.

Kullenberg, G.E.B. [1976]:
"On vertical mixing and the energy transfer from the wind
to the water". Tellus, Vol. 28, pp. 159-165.

Kullenberg, G.E.B. [1977a]:
"Entrainment velocity in natural stratified shear flow".
Estuarine and Coastal Marine Science, Vol. 5. pp 329-338.

Kullenberg, G.E.B. [1977b]:
"Observations of the mixing in the Baltic thermo- and
halocline layers". Tellus, Vol. 29, pp. 572-587.

Laufer, J. [1975]:
"New trends in experimental turbulence research". Annual
Review of Fluid Mechanics, Vol. 7, pp. 307-326.

Löfquist, K. [1960]:
   "Flow and stress near an interface between stratified
   liquids". Physics of Fluids, Vol. 3, pp. 158-175.

Møller, J.S. [1984]:
   "Hydrodynamics of an Arctic fjord". Technical University
   of Denmark, Institute of Hydrodynamics and Hydraulic
   Engineering, Series Paper 34, 197 pp.

Ottesen Hansen, N.-E. [1975]:
   "Entrainment in two-layered flows". Technical University
   of Denmark, Institute of Hydrodynamics and Hydraulic
   Engineering, Series Paper 7.

Ottesen Hansen, N.-E. [1978]:
   "Mixing processes in Lakes". Nordic Hydrology, Vol. 9,
   pp 57-74.

Pickard, G.L. [1961]:
   "Oceanographic features of inlets in the British Columbia
   Mainland Coast". Journal of the Fisheries Research Board
   of Canada, Vol. 18, pp. 907-999.

Pritchard. D.W. [1967]:
   "Observations of circulation in coastal plain estuaries".
   In "Estuaries" (G.H. Lauff, ed.). AAAS Publ. No. 83,
   Washington, D.C.

Roshko, A. [1976]:
   "Structure of turbulent shear flows: A new look".
   (A Dryden Research Lecture). AIAA Journal, Vol. 14.

Saelen, O.H. [1967]:
   "Some features of the hydrography of Norwegian fjords".
   In: Estuaries. Proceedings of a conference, Jekyll
   Island, Georgia 1964. Ed. by G.H. Lauff, Washington D.C.
   American Association for the Advancement of Science,
   Publication No. 83, pp. 63-70.

Sehested, I., D.H.I. report [1982]:
   "A note on entrainment in surface jets discharging into a
   flowing recipient and exposed to the wind". [in Danish].
   Danish Hydraulic Institute, DK-2970, Hørsholm, Denmark.

Sorensen, R.M. and Spencer, E.B. [1971]:
   "Two-dimensional wind setup of oil on water". American
   Society of Civil Engineers, Proc., Vol. 97, WW3,
   pp. 517-530.

Stucchi, D. and Farmer, D.M. [1976]:
   "Deep water exchange in Rupert-Holberg Inlet". Institute
   of Ocean Sciences, Patricia Bay, Victoria B.C., Pacific
   Marine Science Report 76-10.

Svendsen, H. [1976]
"Some hydrographic observations of the Sørfjord during 1972". Freshwater on the Sea. Symposium on the influcence of fresh-water outflow on biological processes in fjords and coastal waters, 1974, Geilo, Norway. The Association of Norwegian Oceanographers.

Tesaker, E. [1973]:
"Horizontal cross-flow temperature gradients in a lake due to Coriolis' force". Hydrology of Lakes. Symposium, Helsinki. International Association of Hydrological Sciences, IAHS-AISH Publication No. 109, pp. 72-78.

Tully, J.P. [1949]:
"Oceanography and prediction of pulp mill pollution in Alberni Inlet". Fisheries Research Board of Canada, Bulletin 83, 169 pp.

Turner, J.S. [1973]:
"Buoyancy effects in fluids". Cambridge University Press, 367 pp.

UNESCO [1981a]:
"Unesco technical papers in marine science". No. 37, UNESCO.

UNESCO [1981b]:
"Unesco technical papers in marine science". No. 38, UNESCO.

Wilkinson, D.L. [1970]:
"Studies in density stratified flows". Water Research Laboratory, University of New South Wales, Australia, Report No. 118, 167 pp.

Wilkinson, D.L. [1971]:
"Containment of oil slicks in the St. Lawrence River", Laboratory Technical Report, HY16, National Research Council, Canada.

Winant, C.D. and Browand, F.K. [1974]:
"Vortex pairing: The mechanism of turbulent mixing-layer growth at moderate Reynolds number". Journal of Fluid Mechanics, Vol. 63, pp. 237-255.

Wiuff, R. [1977]:
"Experiments on the surface buoyant jet". Technical University of Denmark, Institute of Hydrodynamics and Hydraulic Engineering, Series Paper 16, 168 pp.

PART III

CASE STUDIES

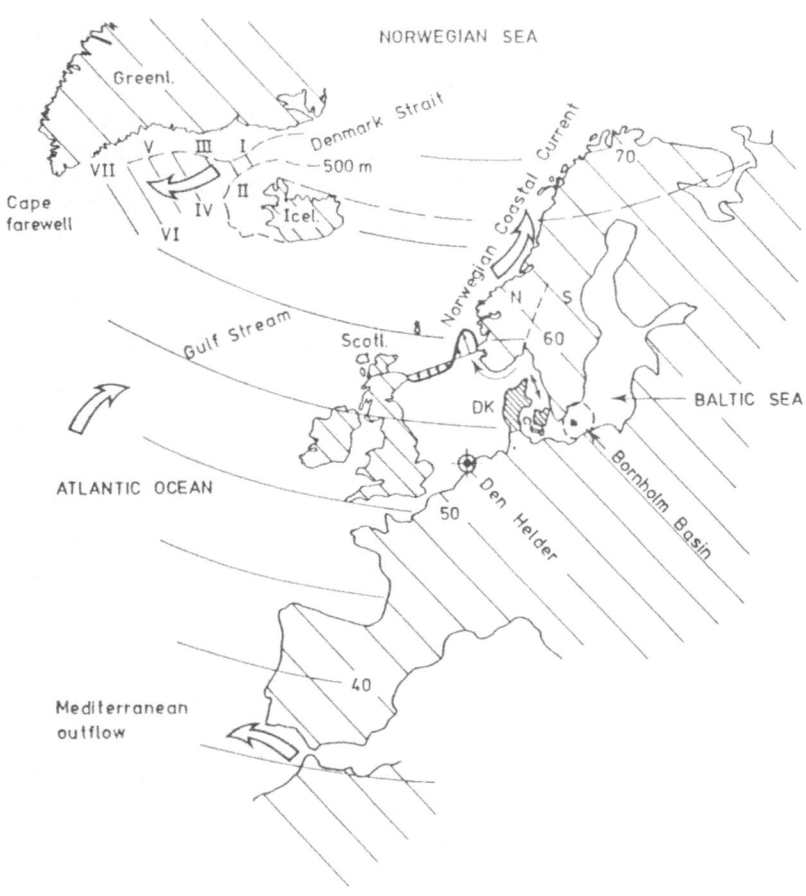

13. COMMENTS ON PART III

In the reprints attached we have tried to give a general picture of what kind of problems within stratified flow which can be solved on the basis of knowledge gained in the present lecture notes. The cases are all an outgrowth of the author's research contributions (except for the one by Engelund [1973]). Paper number 1, 7 and 8 were initiated by the contact with Greenex A/S, who runs a mining activity in the Affarlikassaa Fjord in West Greenland. Contribution number 2 was partly a documentation of the authors bulk flux Richardson number theory (applicable in the whole range of bed slopes from horizontal to vertical and from Reynolds' numbers ranging from 10^4 to 5×10^7), and partly because dense bottom currents are encountered in many geophysical phenomena. Paper number 4 was an outgrowth of a test of the newly built laboratory flume for density stratified flows (paper number 9). The experiments yield entrainment values in free convection in a range not reported on in the literature before, and furthermore, in a range close to the conditions prevailing in nature. Paper number 5 illustrates how to handle a complex estuary system influenced by man's activity. Paper number 6 illustrates one way of treating surface buoyant jets.

The examples given are but a few of the problems which can be handled. On the other hand, a great number of unexplored problems still remains to be solved before our level of knowledge within stratified flows reaches the level we have achieved within ordinary open channel flow. This fact calls especially for new laboratory measurements, see paper number 9.

Prog. Rep. 57, pp. 33-42, November 1982   1
Inst. Hydrodyn. and Hydraulic Engrg.
Tech. Univ. Denmark

4. SEDIMENT LADEN BUOYANCY JETS

by Hans J. Henriksen, Henning Haar   and

Fl. Bo Pedersen*

INTRODUCTION

The discharge of tailing from mining activities and similar discharges of sediment laden jets have a great impact on
the environment, such as local deposition of sediment and spreading of dissolved heavy metals. The keys to the solution of sediment laden jets are the dilution, i.e. the entrainment of ambient water into the jet, and the sediment fall-out from the
jet. Based on a large number of laboratory experiments the rate
of entrainment is determined, as well as the condition for incipient sediment fall-out. Further, a simple model for the sediment fall-out along the path line of the jet has been outlined
and checked upon measurements.

THEORETICAL BACKGROUND

In the design of ocean fall-out diffusers one is mostly
interested in the variations of the gross properties along the
path line. Hence, an integral approach is commonly used, assuming
that the velocity and density profiles are Gaussian across the
jet. Two main difficulties arise in the analysis of sediment
laden jets, namely the rate of entrainment and the rate of sediment fall-out, which both influence the rate of growth and
the buoyancy (or concentration) along the path line. The rate
of entrainment and sediment fall-out have been investigated in
a laboratory experimental setup.

For a stationary buoyant jet/plume without sediment fall-
out, discharging into a homogeneous recipient, the following
conservation equations apply:

horizontal momentum

$$\frac{d}{ds}(\rho\alpha'QV\cos\phi) = 0$$

---

* Abstract of a master thesis (in Danish) by Henriksen and Haar
with Fl. Bo Pedersen as supervisor.

vertical momentum

$$\frac{d}{ds}(\rho\alpha'QV\sin\phi) = \rho\Delta g\pi r^2$$

continuity

$$\frac{dQ}{ds} = 2\pi rV_E$$

buoyancy flux

$$\frac{d\Delta Q}{ds} = 0$$

where ρ = density of mass

α' = velocity distribution coefficient (= 1.56)

Q = discharge

V = mean velocity (= $Q/\pi r^2$)

Δ = mass deficit = $(\rho_{jet} - \rho_{ambient}/\rho_{ambient})$

g = acceleration of gravity

V_E = entrainment velocity = EV

E = proportionality constant determined in the experiments (= 0.07)

r = a local characteristic radius

ϕ = the local inclination of the jet

The conservation equations combined with the geometrical relationships (see figure 1):

$$\frac{dx}{ds} = \cos\phi \qquad \frac{dy}{ds} = \sin\phi$$

can be transformed to a system of five dimensionless differential equations in the variables r/r_0, V/V_0, x/r_0 and y/r_0, where index 0 refers to the outlet conditions. By assuming the length of the zone of flow establishment equal to 12 r_0, the equations have been solved numerically by use of the IMSL-library routine DREBS. The solutions are shown in figures 2 and 3 respectively, where the entrainment ratio E (E = 0.07) has been used as parameter.

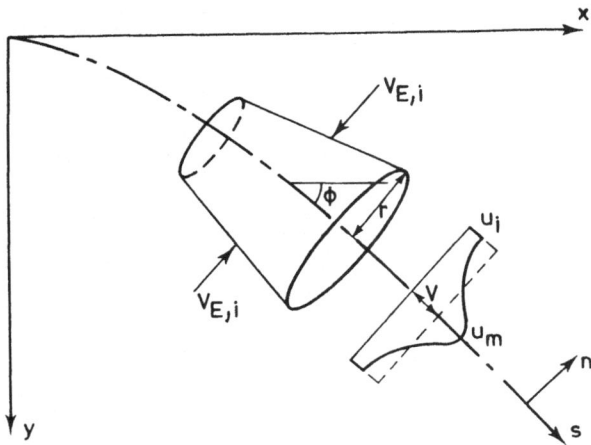

Fig. 1   Definition sketch

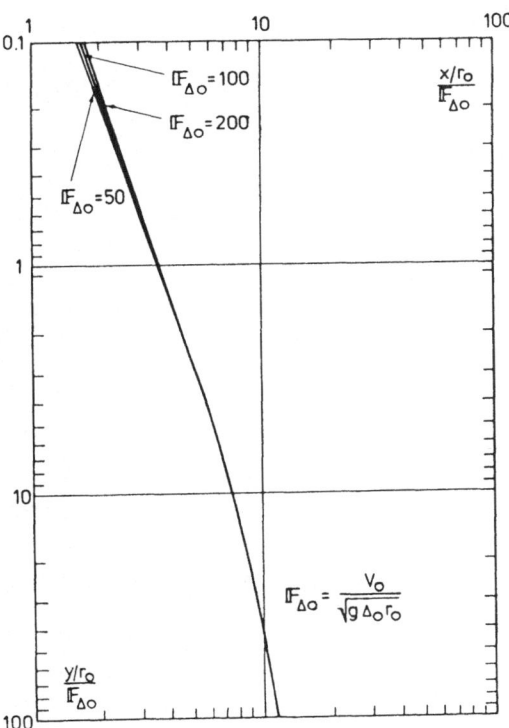

*Fig. 2   Dimensionless pathline for a horizontal
discharged jet. (E = 0.07) (α = 0.057)*

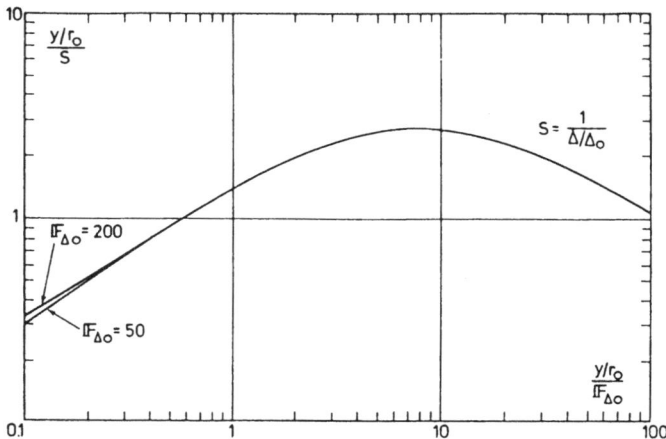

Fig. 3   Mean dilution as function of the depth
         (horizontally discharges jet) (E = 0.07)

EXPERIMENTAL SETUP

General

The main objectives of our experiments are first to determine the entrainment ratio $E = V_E/V$ as a function of the densimetric Froude number

$$\text{IF}_\Delta = \frac{V}{\sqrt{\Delta gr}}$$

and as a function of the concentration (by volume and/or by weight) of the sediment, and secondly to establish the conditions for sediment fall-out from the jet.

The entrainment ratio was determined by a series of experiments with a vertically discharged jet/plume. With the ratio E known, the next series of experiments with horizontally discharged jets yields the conditions for sediment fall-out, i.e. the point of incipient fall-out and the total rate of fall-out.

The experiments were performed in a 8 m section of a 3 m wide and 0.8 m deep flume filled with homogeneous water. The diffuser (radius $r_0 = 2.5 \times 10^{-3}$ m) was fed from a constant head tank, in which the bottom was arranged as a labyrinth-system of small sediment-filled channels. By tracking a suction-head with a constant speed through the labyrinth-system a constant rate of sediment and water was established.

Three different types of sediment were used

$$\text{sand, settling velocity} \quad = \quad \begin{array}{l} 1.58 \times 10^{-2} \text{ m/s} \\ 2.19 \times 10^{-2} \text{ m/s} \end{array}$$

plastic beads, settling velocity $= 1.00 \times 10^{-2}$ m/s

hyperit, settling velocity $\quad = 6.35 \times 10^{-2}$ m/s

The range of variation of the other parameters were

volume concentration (c): 0 - 15%

outlet velocity (V_0): 1.9 - 3.2 m/s

which yields the range of the outlet densimetric Froude number

$$\mathbb{F}_{\Delta,0} = \frac{V_0}{\sqrt{\Delta_0 g r_0}} : 30 - 400$$

and the outlet Reynolds' number

$$\mathbb{R}e_0 = \frac{V_0 \times r_0}{\nu} : 5 \times 10^3 - 8 \times 10^3$$

Vertically discharged jets (VDJ)

In the VDJ no sediment fall-out takes place, and hence our theoretical solution applies. The growth rate of the jet/plume is uniquely related to the rate of entrainment, and hence this can simply be determined by comparing the observed jet/plume with the theoretical. To this end, short time (1/125 seconds) exposure photographs of the VDJ were compared with the theoretical solution for three different E-values (0.06, 0.08 and 0.10). The well-known waviness of the VDJ is obvious from the photographs (see figure 4), and as illustrated in figure 5 the Eulerian spreading is larger than the Lagrangian. This effect has often been overlooked and therefore given rise to an incorrect interpretation of laboratory data on entrainment, for instance that the entrainment rate for plumes (large waviness) should be larger than for jets (small waviness), which has been shown by Bo Pedersen [1980] to be an incorrect statement.

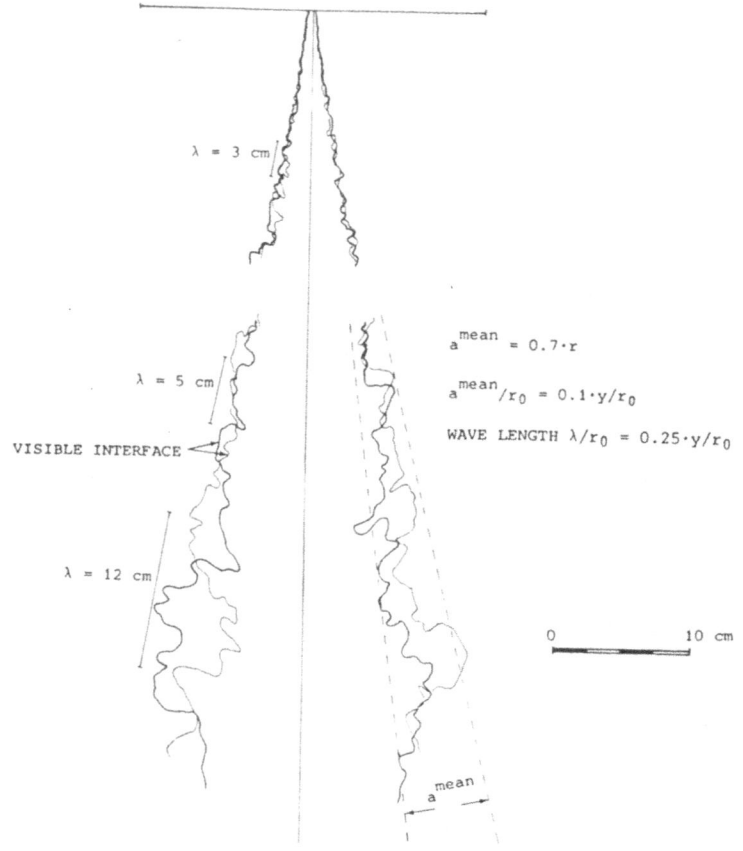

$$a^{mean} = 0.7 \cdot r$$

$$a^{mean}/r_0 = 0.1 \cdot y/r_0$$

WAVE LENGTH $\lambda/r_0 = 0.25 \cdot y/r_0$

Fig. 4   Vertically discharged jet. Sketch of visible inter-
face as observed on the photograph

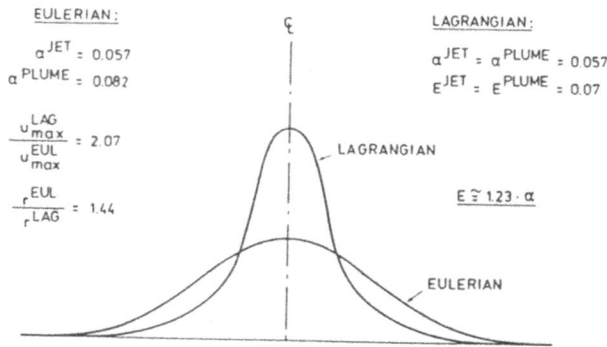

EULERIAN:

$$\alpha^{JET} = 0.057$$
$$\alpha^{PLUME} = 0.082$$

$$\frac{u_{max}^{LAG}}{u_{max}^{EUL}} = 2.07$$

$$\frac{r^{EUL}}{r^{LAG}} = 1.44$$

LAGRANGIAN:

$$\alpha^{JET} = \alpha^{PLUME} = 0.057$$
$$E^{JET} = E^{PLUME} = 0.07$$

$$E \simeq 1.23 \cdot \alpha$$

Fig. 5   The Lagrangian and the Eulerian velocity-profile

Based on our 25 *Lagrangian* experiments on VDJ a constant value of the entrainment ratio was determined

$$E = 0.07 \overset{+}{-} 0.01$$

valid for jets as well as for plumes ($2.4 < \mathbb{F}_\Delta < 200$). No systematic dependence of E on $\mathbb{F}_\Delta$ and/or on the volume concentration was observed.

The entrainment ratio has often been related to the maximum velocity U_m by the equation

$$\frac{dQ}{ds} = \frac{d}{ds}(\pi U_m b^2) = 2\pi\alpha b U_m$$

where b is the nominal radius of the jet ($= \sqrt{2}\sigma$). The so defined entrainment ratio α is proportional to E, and hence

$$\alpha = \sqrt{\frac{b}{r}}\, E = 0.057 \overset{+}{-} 0.008$$

in agreement with the often quoted value for pure jets.

Horizontally discharged jets (HDJ)

In figure 6 the visible boundary of one of the 38 experiments on HDJ is shown. The upper boundary is sharp and well defined, and so is the incipient point of sediment fall-out. Further downstream an intermittent sediment fall-out in diffused streaks is observed.

The point of incipient sediment fall-out turned out to be the location where the settling velocity of the sediment became larger than the local entrainment velocity, i.e. sediment fall-out condition: $W \cos\phi \overset{>}{-} V_E$.

In order to test a simple model for the sediment fall-out, the sediment was collected at the bottom of the tank. In figure 7 an example of the distribution of settled sediment is shown. This has been cross-sectionally integrated to give the one-dimensional longitudinal distribution which is shown. By integration this can easily be transformed to the shown mass-curves (M) for the sediment retained in the jet (relative to the initial value). This M-value is equal to the ratio of the

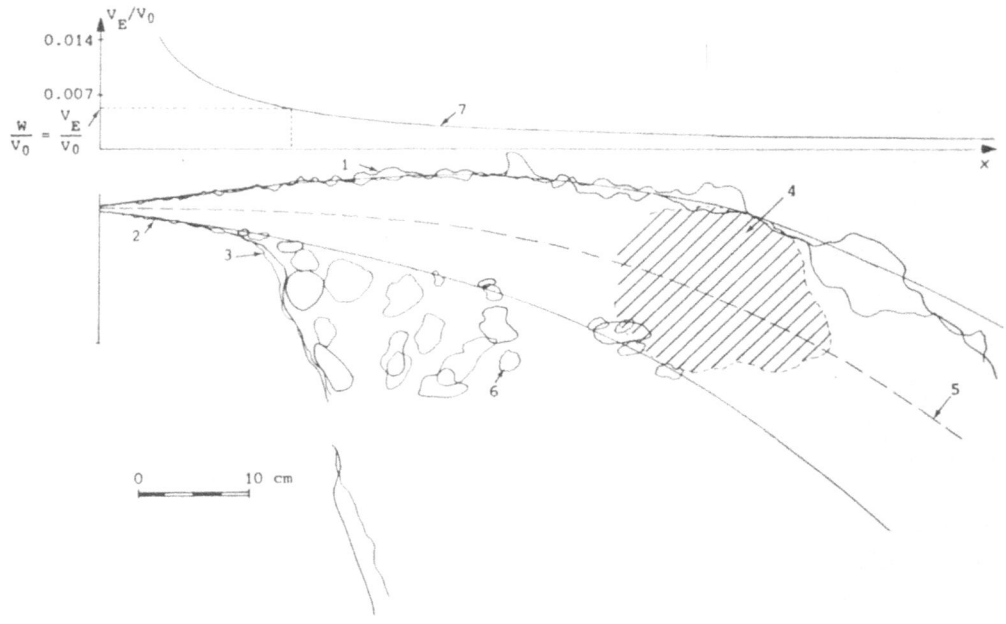

Fig. 6   *Visible boundaries of the horizontally discharged*
          jet

       1) *upper boundary*
       2) *lower boundary*
       3) *incipient sediment fall-out*
       4) *a dispersed pulse-injected dye*
       5) *theoretical center path-line*
       6) *low-concentration holes*
       7) *local dimensionless entrainment velocity*

local buoyancy flux to the outlet buoyancy flux.

In the simple model we assume that the mean concentration is proportional to the concentration at the lower boundary of the jet/plume, i.e. that the concentration profiles at-
tain a state of similarity within a short distance. Obviously,
the concentration distribution depends on the relative strength
of the fall velocity W to the turbulent velocity. Based on
curve-fitting of the observed to the theoretical fall-out model
the following relationship was obtained

$$c_{boundary}/c_{mean} = \frac{600 \ W \ \cos\phi}{\mathbb{F}_{\Delta,0} \times V}$$

The associated net-fall-out velocity is $V_{out} = W \cos\phi - V_E$,
i.e. corrected for the inwards directed entrainment.

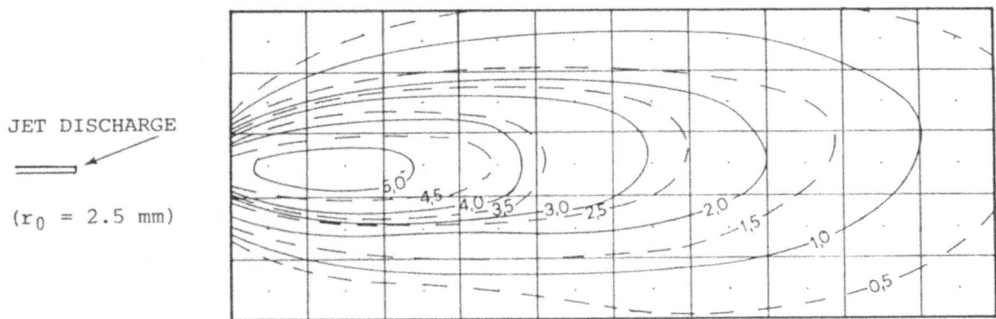

a) *Distribution of settled sediment (number = percent of totally settled sediment)*

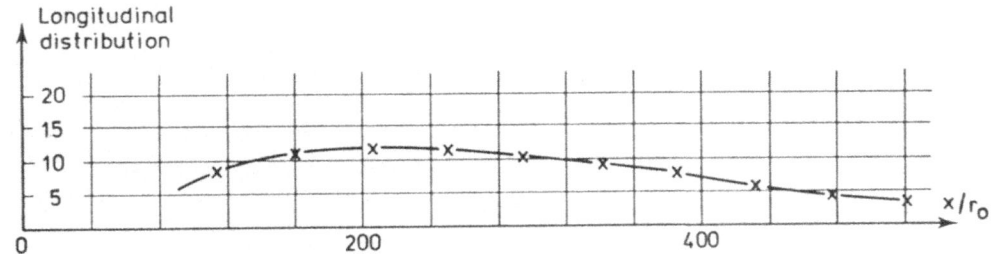

b) *One-dimensional longitudinal distribution of settled sediment*

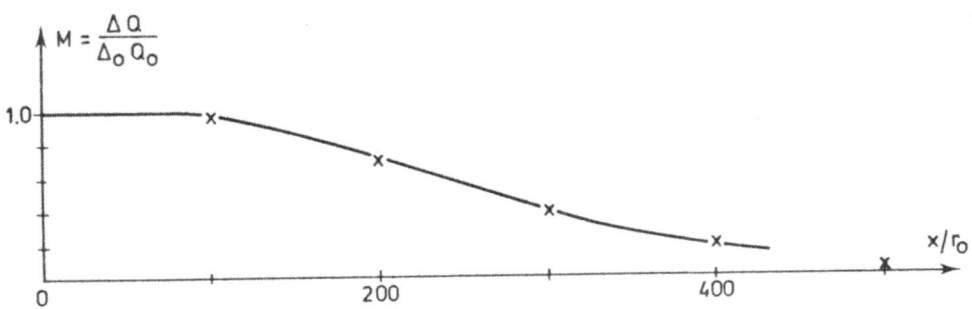

c) *Mass-curves for the sediment retained in the jet. M is equal to the ratio of the local buoyancy flux to the outlet buoyancy flux*

Fig. 7

CONCLUSION

Based on a large number of experiments with vertically discharged sediment laden jets, a constant rate of entrainment $E = V_E/V = 0.07$ (alternatively $\alpha = 0.057$ (see text)) was found, independent of the Froude number and the concentration of sediment. This constant rate of entrainment is valid for the Lagrangian spreading. In an Eulerian framework a spurious spreading is present, which has often led to the *incorrect* conclusion that the rate of entrainment for plumes should be larger than the one for jets. The rate of entrainment $\alpha = 0.057$ is in agreement with often quoted values for pure jets (Lagrangian equal to Eulerian as no waviness is present for pure jets).

The point of incipient sediment fall-out corresponds to the point where the settling velocity exceeds the local entrainment velocity.

The sediment fall-out can be calculated by a simple model, which relates the bottom boundary sediment concentration to the cross-sectional mean concentration.

References:

Bo Pedersen, F., 1980. A monograph on Turbulent Entrainment and Friction in Two-layer Stratified Flow. Series Paper 25, Institute of Hydrodynamics and Hydraulic Engineering, Technical University of Denmark.

Henriksen, H.J. and Haar, H., 1982. Sedimentførende stråler. Master Thesis (in Danish). Supervisor Bo Pedersen, F.

AUGUST 1980

JOURNAL OF THE HYDRAULICS DIVISION

DENSE BOTTOM CURRENTS IN ROTATING OCEAN

By Flemming Bo Pedersen[1]

INTRODUCTION

A dense bottom current is the flow created by a source of mass, momentum, and buoyancy flowing into an ambient fluid in such a way that the flow is bounded by the fixed wall and the interface. The dense bottom currents are primarily driven by buoyancy forces (reduced gravity).

All oceans, all estuaries, and nearly all lakes and reservoirs that receive water from a river have dense bottom currents. In the ocean and in the estuaries the difference in salinity often plays the most important role in creating the forcing excess density, although temperature and turbidity differences can be present too. In lakes and reservoirs the dense bottom currents are initiated by natural or artificial discharge of a fluid that is heavier than the fluid in the recipient, either due to temperature or to turbidity differences.

In the ocean and in the estuaries, the dense bottom currents normally play a crucial role for the ecologic life by being the source of oxygen supply, e.g., in the cases of the Denmark Strait Overflow (considered later) and of the deep-water renewal in fiords. In a previous paper (3) the writer has treated the two-dimensional, fully developed turbulent dense bottom currents in a homogeneous ambient fluid at rest (except for the flow induced by the density current). As the main findings concerning the entrainment and the friction shall be utilized here, a brief summary on nonrotating dense bottom currents is expedient.

NONROTATING DENSE BOTTOM CURRENTS

Dense bottom currents are normally dominated by the buoyancy (gravity) as the driving force and the friction (at the wall and at the interface) as the balancing force. This balance is rapidly reached, which means that the current on a floor with a constant slope very soon reaches a state of equilibrium with

---

Note.—Discussion open until January 1, 1981. To extend the closing date one month, a written request must be filed with the Manager of Technical and Professional Publications, ASCE. This paper is part of the copyrighted Journal of the Hydraulics Division, Proceedings of the American Society of Civil Engineers, Vol. 106, No. HY8, August, 1980. Manuscript was submitted for review for possible publication on August 21, 1979.

[1]Research Engr., Inst. of Hydrodynamics and Hydr. Engrg., Tech. Univ. of Denmark, Lyngby, Denmark.

a constant densimetric Froude number squared F_Δ^2:

$$F_\Delta^2 = \frac{V^2}{\Delta gy} \qquad \ldots \ldots \ldots \ldots \ldots \ldots \ldots \ldots \ldots \ldots (1)$$

in which V = depth average velocity; Δ = dimensionless mass difference between the current and the ambient fluid ($\Delta << 1$); g = acceleration of gravity; and y = depth of the current.

In the limit of infinite slope, i.e. a vertical wall, the bottom current is transferred to a (half-) falling plume. Accordingly, dense bottom currents in the equilibrium state are associated with densimetric Froude numbers in the range of

$$0 < F_\Delta \leq F_{\Delta,plume} = 2 \qquad \ldots \ldots \ldots \ldots \ldots \ldots \ldots \ldots \ldots (2)$$

As $F_{\Delta,plume}$ is well within the supercritical flow range, the entrainment process for dense bottom currents on a steeply sloping bottom is associated with vortex engulfing at the interface, while the density currents on a slightly sloping floor have an entrainment caused by interfacial wave breaking. A diagnostic equation for the entrainment was derived by making use of the new bulk flux Richardson number F_f^T which, very briefly, is the efficiency of the mixing process, i.e., the ratio of the gain in potential as well as in turbulent kinetic energy to the energy available (the production).

By applying the depth integrated work energy equation to the entraining current, a relation between the bed slope I_0 and the bulk Richardson number R_f was obtained as shown in Fig. 1, where experimental and field data are shown as well. Accordingly, by making use of the new bulk flux Richardson number R_f^T, an entrainment function was obtained that is shown in Fig. 2 in which the laboratory and field measurements available have been plotted too. The jumps in the curves in Fig. 1 and Fig. 2 are due to the drastic changes in the velocity profiles and in the bulk flux Richardson number, respectively, that occur in the transition from the sub- to the supercritical flow regime.

By relating the entrainment velocity V_E to the bed slope I_0 a very simple relation is obtained

$$\frac{V_E}{V} \simeq 0.072 \, I_0; \quad F_\Delta < F_{\Delta,cr} \qquad \ldots \ldots \ldots \ldots \ldots \ldots \ldots \ldots (3)$$

that is valid for subcritical flows independently of the friction factor. The bed slope I_0 can be related to the boundary shear stresses as

$$I_0 = \frac{f}{2} F_\Delta^2 = \frac{f}{2} \frac{V^2}{\Delta gy} = \frac{\frac{(\tau_i + \tau_w)}{\rho}}{\Delta gy} \quad \text{for} \quad I_0 \leq 10^{-2} \qquad \ldots \ldots \ldots \ldots (4)$$

which is the well-known relation applicable for nonentraining density currents and ordinary open channel flow ($\Delta = 1$). Eq. 4 expresses the force balance between the flow direction component of the gravity and the friction. It is emphasized that this relationship is applicable only for gentle slopes ($I_0 \leq 10^{-2}$).

For engineering purposes the very simple entrainment function, Eq. 3, can be used as a first approximation for the whole range of bed slopes. This has the very great advantage that density currents influenced by the Coriolis force

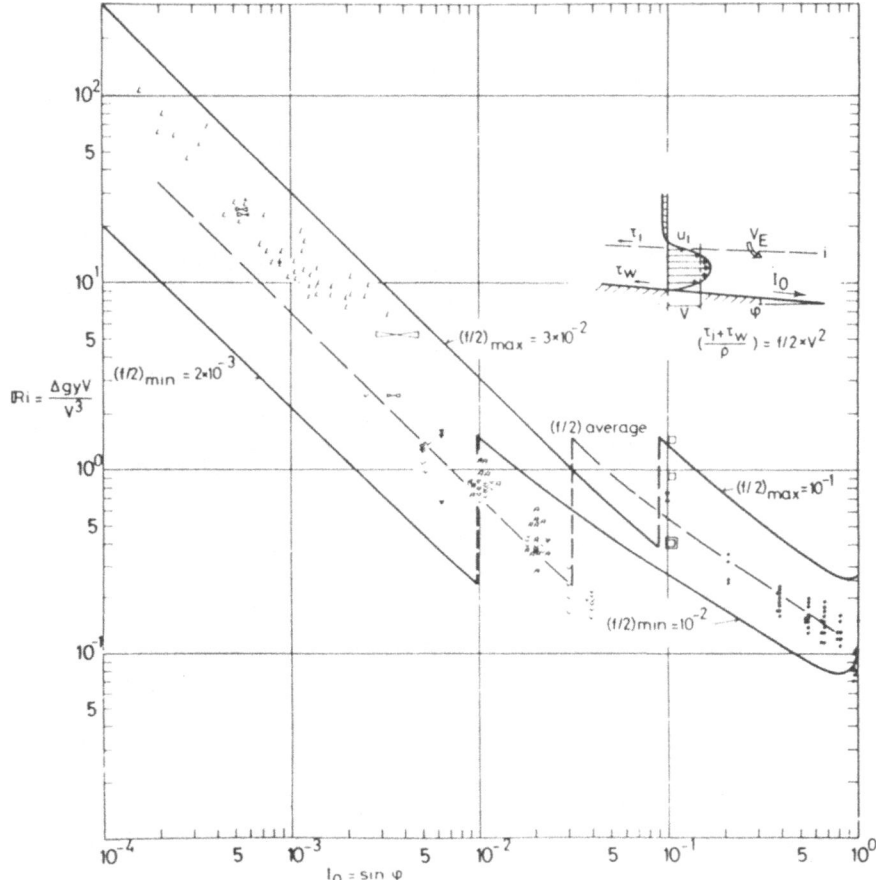

The points are based on data referred by:

v   [9] (Laboratory experiments).

.   [5] (Laboratory experiments).

□   [4] (Deep water renewal of Loch Etive Fjord).

⧗   [6], average of 107 experiments (mining engineering).

A   [1] (Laboratory turbidity currents).

▷◁ [10] and [12] (Dense bottom current in the Denmark Strait).

L   [8] (Laboratory measurements). (The depth y is here R =
    hydraulic radius).

▼   [11] (Laboratory measurements). (y is here = R = hydraulic
    radius).

**FIG. 1.—Laboratory and Field Data Compared with Extreme Limits to Bottom Slope
I_0 Versus Richardson Number Ri (Based on Depth Integrated Work Energy Equation)**

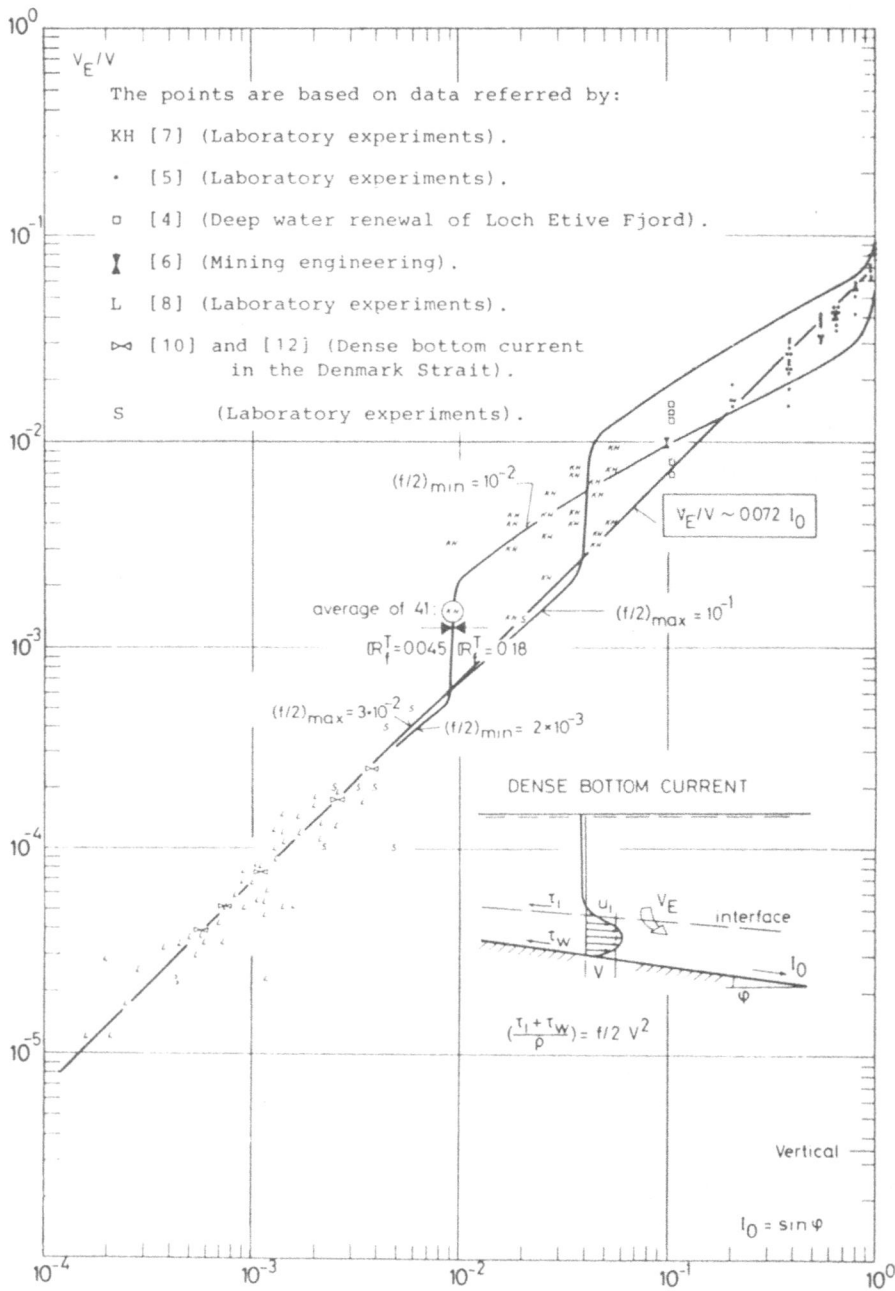

FIG. 2.—Laboratory and Field Data on Entrainment Compared With Theory (Lowest and Highest Entrainment Rates Are Drawn, Based on Estimate of Natural Range of Variation of Friction Coefficient $f/2$)

(that are the most common currents in nature) may be treated by means of a dimensionless length scale l defined by

$$l = \int_0^s \frac{I_0(s)}{y(s)} \, ds \quad \ldots \ldots \ldots \ldots \ldots \ldots \ldots \ldots \ldots \ldots \ldots (5)$$

The use of l makes it possible to overcome the hurdle of the bottom current following a path line, s, with variable bed slope, as shall be demonstrated.

ROTATING DENSE BOTTOM CURRENTS

When a dense bottom current flows into a lighter homogeneous ocean on a plane, it experiences an ever decreasing slope as the Coriolis force continuously deflects the path line (see Fig. 3). Thus the state of an equilibrium densimetric Froude number as observed in nonrotating currents is not present in the ocean.

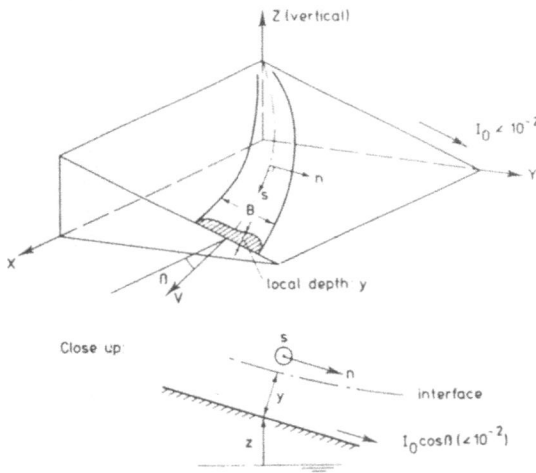

FIG. 3.—Three-Dimensional Dense Bottom Current on Gently Sloping Plane in Rotating Ocean in Northern Hemisphere (Distorted Scale)

On the other hand, if the plane is gently sloping ($I_0 < 10^{-2}$, see Fig. 3), a local balance can be assumed to exist between the flow direction component of the acceleration of gravity, the friction to which the flow is exposed at the bottom, and the interface respectively, with the gentle slope making the momentum terms negligible. Thus, with reference to the preceding findings (Eq. 4) the flow direction component of the equation of movement yields (see Fig. 3):

$$\frac{\partial}{\partial s}(z + y) \simeq -\left(\frac{f}{2}\right)\frac{V^2}{\Delta\, gy} = -\left(\frac{f}{2}\right)F_\Delta^2 \quad \ldots \ldots \ldots \ldots (6)$$

Accordingly, the lateral force balance is between the lateral component of the acceleration of gravity and the Coriolis force, as the ratio of the centrifugal force to the Coriolis force

$$\frac{\left(\dfrac{\rho V^2}{gr}\right)}{\left(\dfrac{\rho f_c V}{g}\right)} = \frac{V}{f_c r} \quad \dots\dots\dots\dots\dots\dots\dots\dots (7)$$

will be small in the general case (r = the radius of curvature; and $f_c = 2\omega$ $\sin \Phi$ is the Coriolis parameter). Thus, the streamwise velocity V can be assumed to be in a geostrophic balance, i.e.

$$\frac{\partial}{\partial n}(z + y) \simeq -\frac{f_c V}{\Delta g} \quad \dots\dots\dots\dots\dots\dots\dots (8)$$

The total discharge of the current

$$Q = \int_0^B V y \, dn \quad \dots\dots\dots\dots\dots\dots\dots\dots (9)$$

increases slightly in the flow direction at a rate determined by the total entrainment into the current, i.e.

$$\frac{\partial Q}{\partial s} = \int_0^B V_E \, dn \quad \dots\dots\dots\dots\dots\dots\dots (10)$$

in which the entrainment velocity V_E is related to the local slope of the current (see Eq. 3):

$$\frac{V_E}{V} = 0.072 \, I_0 \sin \beta = 0.072 \frac{f}{2} \frac{V^2}{\Delta g y} \quad \dots\dots\dots\dots (11)$$

In Eq. 11 the geometric relationship

$$\frac{\partial}{\partial s}(z + y) \simeq -I_0 \sin \beta \quad \dots\dots\dots\dots\dots\dots (12)$$

has been utilized where it has been assumed that the depth variation is moderate ($\partial y << \partial z$). The lateral adversary to Eq. 12

$$\frac{\partial}{\partial n}(z + y) \simeq -I_0 \cos \beta \quad \dots\dots\dots\dots\dots\dots (13)$$

shall be applied as well.

For the sake of simplicity the ambient fluid has been taken to be homogeneous which implies that the mass deficit flux

$$A = \int_0^B (\Delta V y) \, dn \quad \dots\dots\dots\dots\dots\dots\dots (14)$$

is conserved in the flow direction that can be expressed as

$$\frac{\partial A}{\partial s} = 0 \quad \dots\dots\dots\dots\dots\dots\dots\dots\dots (15)$$

The preceding set of partial differential equations can be solved only if the existence of similarity solutions is assumed. This is, on the other hand, a probable assumption when a current on a plane is concerned.

Combining Eqs. 6, 8, 12, and Eq. 13 yields

$$\left(\frac{f}{2}\right)^2 F_\Delta^4 + \left(\frac{f_c V}{\Delta g}\right)^2 = I_0^2 (\sin^2 \beta + \cos^2 \beta) = I_0^2 \quad \ldots \ldots \ldots \ldots \quad (16)$$

which can be solved with respect to the Coriolis force term

$$\frac{f_c V}{\Delta g} = I_0 \sqrt{1 - \frac{\left(\frac{f}{2}\right)^2 F_\Delta^4}{I_0^2}} \simeq I_0 \left\{ 1 - \frac{1}{2}\left[\frac{\left(\frac{f}{2}\right) F_\Delta^2}{I_0}\right]^2 \right\} \quad \ldots \ldots \ldots \quad (17)$$

in which the Taylor expansion of the square root can be used for, say

$$\frac{\left(\frac{f}{2}\right) F_\Delta^2}{I_0} = \sin \beta \leq 0.5 \quad \ldots \ldots \ldots \ldots \ldots \ldots \quad (18)$$

which is fulfilled, e.g., for the Denmark Strait Overflow (see Fig. 8). Now Eqs. 6 and 8 are cross-differentiated to yield

$$\frac{\partial}{\partial n}\left[\left(\frac{f}{2}\right) F_\Delta^2\right] = \frac{\partial}{\partial s}\left(\frac{f_c V}{\Delta g}\right). \quad \ldots \ldots \ldots \ldots \quad (19)$$

in which the Coriolis force term can be eliminated by means of Eq. 17

$$\frac{\partial}{\partial n}\left[\left(\frac{f}{2}\right) F_\Delta^2\right] = -\frac{1}{2}\frac{\left(\frac{f}{2}\right)^2}{I_0}\frac{\partial F_\Delta^4}{\partial s} \quad \ldots \ldots \ldots \ldots \quad (20)$$

In order to proceed further the following similarity assumptions are introduced

$$F_\Delta = F(\eta) F(l); \quad V = V(\eta) V(l); \quad \Delta = \Delta(\eta) \Delta(l); \quad y = y(\eta) y(l) \quad \ldots \ldots \quad (21)$$

in which the characteristic properties of the current are assumed to be described as the product of a function of the lateral and the longitudinal dimensionless coordinates respectively

$$d\eta = \frac{dn}{B(l)}; \quad dl = \frac{\left(\frac{f}{2}\right) F(l)^2}{y(l)} ds \quad \ldots \ldots \ldots \ldots \quad (22)$$

Introducing the similarity assumptions in Eq. 20 and separating the variable yields:

$$-\frac{\partial F(\eta)^{-2}}{\partial \eta} = -\frac{1}{2}\frac{\left(\frac{f}{2}\right)^2}{I_0}\frac{B(l)}{y(l)}\frac{\partial F(l)^4}{\partial l} = c \quad \ldots \ldots \ldots \ldots \quad (23)$$

in which $c(\eta) = c(l) = c$ is a constant.

To facilitate further calculations the similarity expressions are introduced into the continuity Eqs. 9 and 10, respectively. Thus

$$Q = \int_0^B V y \, dn = V(l) \, y(l) \, B(l) \quad \ldots \ldots \ldots \ldots \ldots \ldots \quad (24)$$

in which, for convenience, the finite integral

$$\int_0^1 V(\eta) \, y(\eta) \, d\eta = 1 \quad \ldots \ldots \ldots \ldots \ldots \ldots \quad (25)$$

has been chosen to be one, and

$$\frac{dQ}{dl} = 0.072 \, Q \quad \ldots \ldots \ldots \ldots \ldots \ldots \quad (26)$$

in which, accordingly, the finite integral

$$\int_0^1 V(\eta) \, F(\eta)^2 \, d\eta = 1 \quad \ldots \ldots \ldots \ldots \ldots \ldots \quad (27)$$

has been normalized.

Thus, the discharge as a function of the dimensionless distance is simply (Eq. 26 integrated):

$$\frac{Q}{Q(0)} = \exp(0.072 \, l) \quad \ldots \ldots \ldots \ldots \ldots \ldots \quad (28)$$

in which $Q(0)$ = the discharge at $l = s = 0$.

The similarity assumption introduced into the mass deficit flux Eq. 14 yields

$$A = \Delta(l) \, Q \quad \ldots \ldots \ldots \ldots \ldots \ldots \ldots \quad (29)$$

in which the integration constant has been chosen to be

$$\int_0^1 \Delta(\eta) \, V(\eta) \, y(\eta) \, d\eta = 1 \quad \ldots \ldots \ldots \ldots \ldots \quad (30)$$

As the mass deficit flux is conserved (Eq. 15), Eq. 28 yields the mass deficit variation

$$\frac{\Delta(l)}{\Delta(0)} = \exp(-0.072 \, l) \quad \ldots \ldots \ldots \ldots \ldots \quad (31)$$

The streamwise variation of the densimetric Froude number can be evaluated by Eq. 23 provided that the width to depth ratio can be eliminated. The assumption of $\sin \beta$ being small (<0.5) implies that the Coriolis force is nearly insensitive to a variation in η, i.e.

$$\frac{f_c V}{\Delta g} = \frac{f_c V(\eta) \, V(l)}{\Delta(\eta) \, \Delta(l) \, g} = I_0 \cos\beta \simeq I_0 \quad \ldots \ldots \ldots \ldots \quad (32)$$

$$\text{or} \quad \frac{V(l)}{\Delta(l)} \simeq I_0 \frac{g \Delta(\eta)}{f_c V(\eta)} = c_l \quad \ldots \ldots \ldots \ldots \quad (33)$$

in which c_l is a quasi-constant (subject to small variations).

Thus, by means of the definition of the densimetric Froude number

$$F(l)^2 = \frac{V(l)^2}{\Delta(l)\,gy(l)} = \frac{c_l^2}{g}\frac{\Delta(l)}{y(l)} \quad \dots \quad (34)$$

an expression for the depth variation is obtained

$$y(l) = \frac{c_l^2}{F(l)^2}\frac{\Delta(0)\exp(-0.072\,l)}{g} \quad \dots \dots \quad (35)$$

Accordingly, the width B can be eliminated by making use of the equation of continuity, Eq. 24:

$$B = \frac{Q}{V(l)\,y(l)} = \frac{Q_0 g}{c_l^3\,\Delta(0)^2}\,F(l)^2\exp(0.216\,l) \quad \dots \dots \quad (36)$$

in which Eqs. 28, 33, and 35 have been utilized.

Thus, the width to depth ratio is

$$\frac{B}{y(l)} = \frac{Q(0)\,g^2}{c_l^5\,\Delta(0)^3}\,F(l)^4\exp(0.288\,l) \quad \dots \dots \quad (37)$$

which, when inserted in Eq. 23, yields a differential equation in $F(l)$ that can readily be solved to yield

$$\frac{F(l)^2}{F(0)^2} = \exp(-0.072\,l) \quad \dots \dots \dots \quad (38)$$

Now the depth and width variations are readily found by introducing Eq. 38 into Eq. 35 and Eq. 36, respectively

$$y(l) = y(0) \quad \dots \dots \dots \dots \dots \quad (39)$$

$$B(l) = B(0)\exp(0.144\,l) \quad \dots \dots \dots \dots \quad (40)$$

The lateral variation of $F(\eta)^2$ can be found by integrating Eq. 23, which yields

$$F(\eta)^2 = \frac{1}{\left[\dfrac{1}{F(\eta=0)^2}\right] - c\eta} \quad \dots \dots \quad (41)$$

in which the constant c can be evaluated to be (see Eqs. 23, 38, 39, and 40):

$$c = 0.072\left[\frac{\dfrac{f}{2}F(0)^4}{I_0}\right]\left[\frac{B(0)}{y(0)}\right]\frac{f}{2} \quad \dots \dots \quad (42)$$

If $F(\eta=0)^2 = 1$, then the variation in $F(\eta)$ will normally be weak (e.g., for the Denmark Strait Overflow, $c \simeq 10^{-1}$). Therefore the densimetric Froude number can be taken as laterally quasi-constant,

$$F(\eta)^2 \simeq 1 \quad \dots \dots \dots \dots \quad (43)$$

which implies that the lateral depth variation is weak too

$$y(\eta) \simeq 1 \dots \dots \dots \dots \dots \dots \dots \dots \dots \tag{44}$$

(see Eqs. 25 and 27) and, accordingly, that

$$\Delta(\eta) \simeq 1 \quad \dots \dots \dots \dots \dots \dots \dots \dots \tag{45}$$

(see Eq. 30). Since the lateral variation in the Coriolis force was weak too, (see Eq. 33) the lateral velocity variation yields:

$$V(\eta) \simeq 1 \quad \dots \dots \dots \dots \dots \dots \dots \tag{46}$$

In summary, the characteristic properties of the dense bottom current on a gently sloping rotating plane are evaluated to be

$$\frac{F_\Delta^2}{F_\Delta(0)^2} = \frac{\Delta}{\Delta(0)} = \frac{V}{V(0)} = \frac{\sin\beta}{\sin\beta_0} = \exp(-0.072\,l);$$

$$\frac{Q}{Q(0)} = \exp(+0.072\,l); \quad \frac{B}{B(0)} = \exp(+0.144\,l); \quad \frac{y}{y(0)} = 1 . \dots \dots \tag{47}$$

in which the nondimensional streamwise length is defined as

$$l = \int_0^s \frac{I_0 \sin\beta}{y}\, ds \quad \dots \dots \dots \dots \dots \tag{48}$$

This length scale is equal to the change in elevation along the path line of the current, nondimensionalized by the depth, i.e.

$$\frac{Z(0) - Z}{y(0)} = l . \dots \dots \dots \dots \dots \dots \dots \dots \tag{49}$$

The X-Y coordinates of the path line are given by the geometric relationships (see Fig. 3):

$$\frac{\partial Y}{\partial s} = \sin\beta; \quad \frac{\partial X}{\partial s} = \cos\beta \quad \dots \dots \dots \dots \dots \tag{50}$$

which can be solved with respect to l:

$$\frac{dY}{dl} = \frac{y(0)\sin\beta_0 \exp(-0.072\,l)}{\left(\dfrac{f}{2}\right) F(0)^2 \exp(-0.072\,l)} = \frac{y(0)}{I_0} \quad \dots \dots \dots \tag{51}$$

and
$$\frac{dX}{dl} = \frac{y(0)\left[1 - \dfrac{1}{2}\sin^2\beta_0 \exp(-0.144\,l)\right]}{\left(\dfrac{f}{2}\right) F(0)^2 \exp(-0.072\,l)} \quad \dots \dots \dots \tag{52}$$

which yields $\dfrac{I_0 Y}{y(0)} = l$. $\dots \dots \dots \dots \dots \dots$ (53)

and $\dfrac{I_0 X}{y(0)} = \dfrac{1}{0.072 \sin \beta_0} \left\{ \left[\exp(0.072\,l) + \dfrac{1}{2} \sin^2 \beta_0 \exp(-0.072\,l) \right] \right.$

$\left. - \left(1 + \dfrac{1}{2} \sin^2 \beta_0 \right) \right\}$ . . . . (54)

respectively.

DENMARK STRAIT OVERFLOW

Iceland is the "fixed light" to the ocean currents in the northern part of the Atlantic Ocean, (see Fig. 4). The major contribution to the influx of water into the Norwegian Sea is the Norwegian coastal current which passes through

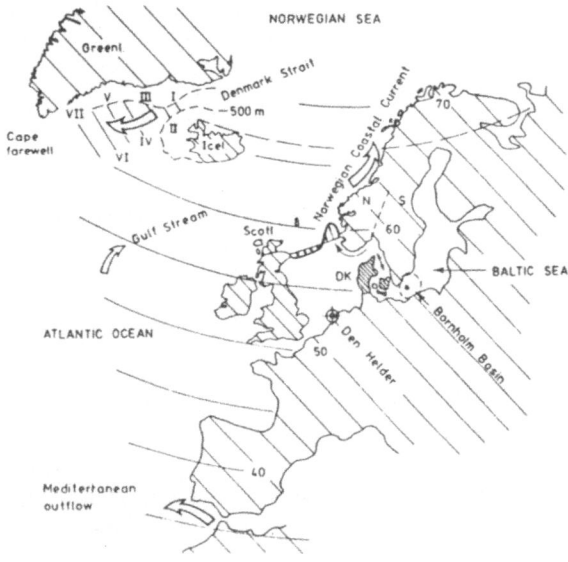

FIG. 4.—Chart of Northeastern Part of Atlantic Ocean

the Iceland-Norway section. In the arctic region there is a sinking of the water due to penetrative convection (cooling). This water (rich in oxygen) sinks to a submarine pool which is dammed up by the shallow ridges from Greenland to Iceland and from Iceland to Scotland. We shall very briefly consider the dense bottom current through the Denmark Strait (named the Denmark Strait Overflow) as a representative of the largest oceanic dense bottom currents.

Fig. 4 shows the location of the Denmark Strait bottom current as traced by hydrographic sections occupied during cruise 0267 of the C.S.S. Hudson out of Bedford Institute of Oceanography, Dartmouth, Nova Scotia, Canada, from January–April, 1967.

In Fig. 5(a) and (b) the cross sections and some typical profiles of potential density, oxygen, and silicates respectively are shown [from Smith (10)]. Smith has treated this current as well as the Mediterranean outflow (see Fig. 4) using a stream tube model that takes the entrainment and the friction into account. By fitting the measured data to his steady-state model, Smith was able to evaluate the entrainment and the friction factors respectively for the current. In Table 1 three sets of data are collected for the current, according to Smith's (10), Worthington's (2), and the writer's findings.

The orders of magnitude of the estimated discharges are nearly identical, but when it comes to the basic physical numbers, namely the entrainment function and the friction factor, the values obtained by Smith's stream tube model are approx one order of magnitude too high.

The basic assumption in the models by Smith and the writer according to which the current takes place on a plane, is not fulfilled at all, and this specific

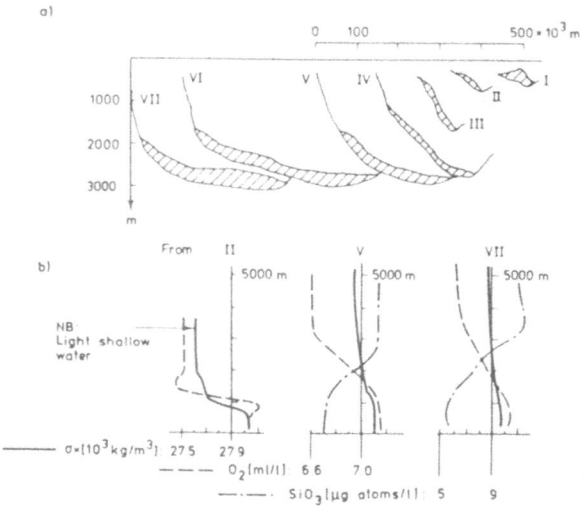

FIG. 5.—Denmark Strait Overflow: (a) Cross Sections; (b) Profiles of Potential Density, σ, Oxygen, O_2, and Silicates, SiO_3 for Typical Stations [Figures after Ref. (10)]

case must, therefore, be treated in a slightly different way. With reference to Fig. 6, the geostrophic balance is written as:

$$\rho f_c V = \frac{\partial p}{\partial x_1} \quad \ldots\ldots\ldots\ldots\ldots\ldots\ldots\ldots\ldots \quad (55)$$

in which $f_c = 2\omega \sin \Phi$ is the Coriolis parameter ($\approx 1.3 \times 10^{-4}$/sec in the present case); V = the velocity in the flow direction; x_1 = horizontal coordinate perpendicular to the flow direction; and p = the excess pressure (i.e., the pressure above the ambient fluid pressure).

To facilitate the use of the geostrophic balance an integration over the cross-sectional area is performed

$$\int_F (\rho f_c V)\, dF = \int_F \left(\frac{\partial p}{\partial x_1}\right) dF \quad \ldots\ldots\ldots\ldots\ldots \quad (56)$$

which yields (see Fig. 6):

$$\rho f_c Q = P = \int_0^B \left[\frac{\Delta \rho g \left(\dfrac{y}{\cos \phi} \right)}{\left(\dfrac{y}{\sin \phi} \right)} \right] y \left(\frac{dx_3}{\sin \phi} \right)$$

$$= \int_0^h \left(\frac{\Delta \rho g y}{\cos \phi} \right) dx_3 = \beta \times \Delta \rho g y h \quad \dots \dots \dots \dots \dots \dots (57)$$

in which the pressure distribution coefficient β is unknown (but assumed to be a constant).

TABLE 1.—Data for Denmark Strait Overflow

Author (1)	$Q_{entrance}$, in cubic meters per second (2)	$Q_{Cape\ Farewell}$, in cubic meters per second (3)	V_E/V (4)	$f/2$ (5)
Smith (10)	1.3×10^6	4.6×10^6	65×10^{-5}	15×10^{-2}
Worthington (12)	4×10^6	10×10^6		
Bo Pedersen (see text)			$0.070\ I_0$*	10^{-2}

*$4 \times 10^{-5} \leq V_E/V \leq 2.5 \times 10^{-4}$.

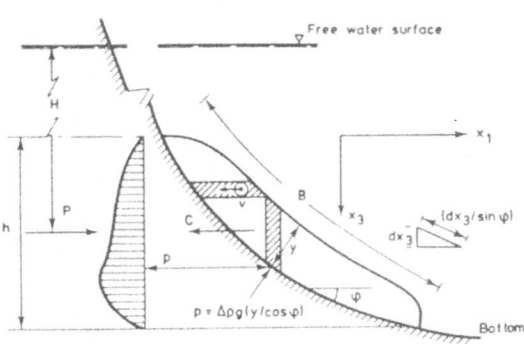

FIG. 6.—Sketch of Southgoing Dense Bottom Current in Northern Hemisphere [Highly Distorted Scale (cos $\phi \simeq 1$)]

This equation can be solved with respect to the discharge Q to give

$$Q = \frac{\beta \Delta g y h}{f_c} \quad \dots \dots \dots \dots \dots \dots (58)$$

By insertion of the constant mass-deficit flux in Eq. 58 an alternative expression for the discharge can be obtained:

$$Q = \sqrt{\frac{\beta (\Delta Q) g}{f_c}} \sqrt{yh} = \text{constant } \sqrt{yh} \quad \dots \dots \dots \dots (59)$$

which has the advantage that the variation in the discharge is related to the overall geometry of the current, which is quite well known (see Fig. 5).

The physical and the geometrical properties of the current have been collected in Table 2 [evaluated by means of the data published by Smith (10)].

When the density deficit Δ and the inverse discharge coefficient $(yh)^{-1/2}$ are made dimensionless by means of the respective values at their arbitrary, but common, origin, they both fit the same exponential curve (see Fig. 7),

TABLE 2.—General Geometric and Physical Properties of Denmark Strait Overflow

Section number (1)	Occupied at (2)	y, in meters (3)	h, in meters (4)	$200\,m/\sqrt{yh}$ [a] (5)	H, in meters (6)	Δs, in meters (7)	$I_0 \sin\beta$ $= \Delta H/\Delta s$ (8)	$\Delta I = \Delta H/y$ (9)	I (10)	$\Delta/3 \times 10^{-4}$ [a] (11)
I	end of January	—	—	—	—				0	—
II	end of January	150	500	0.73	500	118×10^3		3.3 [b]	3.3	0.77 [c]
III	early February	—	--	—	1.000	142×10^3	3.5×10^{-3}	3.3	6.6	0.57
IV	end of March	180	1.600	0.37	2.200	252×10^3	4.8×10^{-1}	6.7	13.3	0.43
V	end of March	200	1.500	0.37	2.700	182×10^3	2.8×10^{-3}	2.6	15.9	0.37
VI	end of March	240	1.300	0.36	2.800	177×10^3	5.6×10^{-4}	0.5	16.4	0.33
VII	early February	400	1.300	0.28	2.900	182×10^3	5.5×10^{-4}	0.3	16.7	0.23

[a] The meaning of the constants will be elucidated in the text.
[b] Can be chosen arbitrarily, as we are only concerned with the flow after Section No. II.
[c] This density excess is not the local value, as the entrainment downstream of Section No. II is not of shallow water (see Fig. 5).

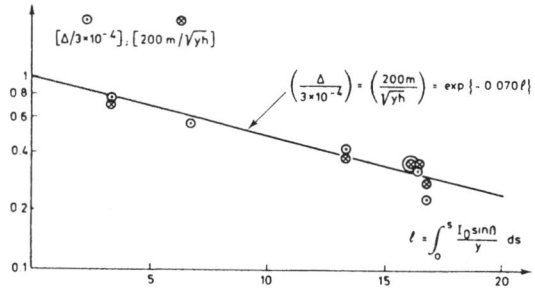

FIG. 7.—Dimensionless Density and Inverse Discharge Respectively, as Function of Dimensionless Distance of Denmark Strait Overflow

with a dimensionless length scale of $14.3 = 0.07^{-1}$, i.e.

$$\left(\frac{\Delta}{3 \times 10^{-4}}\right) = \left(\frac{200\,m}{\sqrt{yh}}\right) = \exp\left(-0.070 \int_0^s \frac{I_0 \sin\beta}{y} ds\right) \quad \dots \dots \dots (60)$$

or $\left(\dfrac{\Delta}{\Delta(0)}\right) = \dfrac{\sqrt{y(0)\,h(0)}}{\sqrt{yh}} = \dfrac{Q(0)}{Q} = \exp\{-0.070\,l\}$ . . . . (61)

The first equality simply states that the mass deficit flux ΔQ is a constant for the current. This is in agreement with the actual, extremely weak stratification in the ambient fluid [see Smith (10)]. From the discharge relationship the following entrainment function can be evaluated

$$\frac{V_E}{V} = 0.070\, I_0 \sin\beta \quad \dots\dots\dots\dots\dots\dots\dots\dots\dots \quad (62)$$

in agreement with the theory by the writer (3) and the observations available, (see Fig. 2). The entrainment coefficient, 0.070, is shown in Ref. 3 to be a constant, 1.6, multiplied by the constant bulk flux Richardson number R_f^T. Consequently, the entrainment value obtained by means of Smith's stream tube model (10) corresponds to a flux Richardson number of about 0.5 that is an order of magnitude higher than the findings of all other researchers.

The discharges in the flow situations reported can be evaluated by Eq. 58

$$Q(0) = \frac{\Delta(0)\, gy(0)\, h(0)}{f_c} = \frac{(3 \times 10^{-4})\, 9.81\,(200)^2\beta}{1.3 \times 10^{-4}} = 0.9 \times 10^6\, \beta\; \text{m}^3/\text{s} \quad (63)$$

which yields for the discharge as a function of distance l:

$$Q = \beta \times 0.9 \times 10^6\, \text{m}^3/\text{s} \exp\,(0.070\, l) \quad \dots\dots\dots\dots\dots \quad (64)$$

The values of the discharges shown in Table 1 can be used for the estimation of the pressure distribution coefficient β

$$\beta\,(\text{Worthington's data}) = 3.5 \quad \dots\dots\dots\dots\dots\dots\dots\dots \quad (65)$$

According to Smith, Worthington used estimates from dynamic computations and neutrally buoyant float measurements to arrive at the transport values

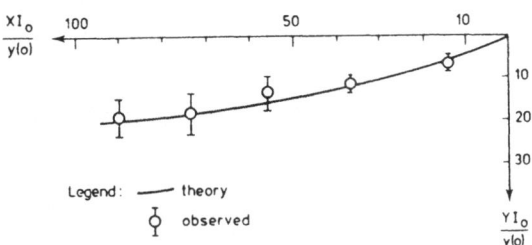

FIG. 8.—Nondimensional Path Line of Dense Bottom Current in Rotating Ocean (sin $\beta_0 = 0.5$) Compared with Observed Path of Stream Axis for Denmark Strait Overflow ($I_0/y(0) = 10^{-4}$); Observed Values After Ref. 10

mentioned. The pressure distribution coefficient $\beta = 3.5$ may not be taken as universal, as it highly depends on the geometry of the slope and on the way in which y, Δ, and h are defined.

The constant mass-deficit flux for the Denmark Strait Overflow therefore is

$$\Delta(0)\, Q(0) = (3 \times 10^{-4})(3.2 \times 10^6) \approx 960\, \text{m}^3/\text{s} \quad \dots\dots\dots\dots \quad (66)$$

The calculated path of stream axis for the current [sin $\beta_0 = 0.5$, $I_0/y(0) = 10^{-4}$/m] is compared with the observed path in Fig. 8, where good agreement is present.

Finally an estimate of the order of magnitude of the friction factor defined by

$$\frac{(\tau_i + \tau_w)}{\rho} = \left(\frac{f}{2}\right) V^2 \simeq \Delta\, g y I_0 \sin\beta \quad\dots\dots\dots\dots\dots\dots\dots \quad (67)$$

yields

$$\frac{f}{2} \approx \frac{\Delta\, g y I_0 \sin\beta}{V^2} \simeq (\Delta Q)\, g\left(\frac{y F^2 I_0 \sin\beta}{Q^3}\right) \quad\dots\dots\dots\dots\dots \quad (68)$$

The calculated values of $f/2$ in Table 3 may be taken only as orders of magnitude. The average value is an order of magnitude less than the value obtained by Smith (10), but, still, it is rather high. This may indicate either that the velocities are too low in the calculations, or that the current is fluctuating above the values estimated by Worthington, or that there really are roughness elements of the order of magnitude of 10 m–50 m high. No further speculation about this matter shall be done, but instead a final comment on the discharge problem will be made. As the bottom water renewal in the northern part of the Atlantic Ocean is crucial to the whole aquatic life of that part of the ocean, it was worthwhile to perform continuous measurements of the potential density,

TABLE 3.—Estimate of Friction Factor ($f/2$) for Denmark Strait Overflow

Section number (1)	l (2)	y, in meters (3)	B, in meters (4)	F, in square meters (5)	$I_0 \sin\beta$ (6)	$(f/2)$ (7)
I	0	—	—	—	—	—
II	3.3	150	75×10^3	11×10^6	3.5×10^{-3}	0.9×10^{-2}
III	6.6	—	—	—	4.8×10^{-3}	—
IV	13.3	180	200×10^3	36×10^6	2.8×10^{-3}	1.6×10^{-2}
V	15.9	200	240×10^3	48×10^6	5.6×10^{-4}	0.8×10^{-2}
VI	16.4	240	400×10^3	96×10^6	5.5×10^{-4}	1.1×10^{-2}
VII	16.7	400	330×10^3	132×10^6		3.3×10^{-2}

oxygen content, etc., just upstream of the overflow where the conditions are much more stable. These measurements would make it possible to obtain a more precise estimate of the quantities transported with the overflow as a function of time. This has been demonstrated for the Bornholm basin which supplies the deep water of the central part of the Baltic (see Fig. 4 and Ref. 2).

SUMMARY

The characteristic features of two-dimensional dense bottom currents have been summarized, mainly based on the findings in Ref. 3.

These findings have been inproved on in establishing the basic equations for the dense entraining bottom current on a plane in a rotating ocean. By assuming the existence of similarity solutions, all the characteristic properties of the dense bottom current can be described as functions of a streamwise length scale l that can be interpreted as the change in elevation along the continuously deflected path line, nondimensionalized by the depth.

Finally, the huge Denmark Strait Overflow has been treated. As the basic assumption of the flow taking place on a plane fails, it became necessary to use a slightly different approach involving the same basic physical assumptions as for the plane current. The properties measured for the Denmark Strait Overflow all confirm the usefulness of the model outlined.

APPENDIX I.—REFERENCES

1. Ashida, K., and Egashira, S., "Basic Study on Turbidity Currents," *Transactions*, Japan Society of Civil Engineers, Vol. 7, 1975, pp. 83–86.
2. Bo Pedersen, F., "On Dense Bottom Currents in the Baltic Deep Water," *Nordic Hydrology*, Vol. 8, No. 5, 1977, pp. 297–316.
3. Bo Pedersen, F., "A Monograph on Turbulent Entrainment and Friction in Two-Layer Stratified Flow," *Series Paper 25*, Institute of Hydrodynamics and Hydraulic Engineering, Technical University of Denmark, Lyngby, Denmark, 1980.
4. Edwards, A., and Edelsten, D. J., "Deep Water Renewal of Loch Etive: A Three Basin Scottish Fjord," *Estuarine and Coastal Marine Science*, Vol. 5, 1977, pp. 575–595.
5. Ellison, T. H., and Turner, J. S., "Turbulent Entrainment in Stratified Flows," *Journal of Fluid Mechanics*, Vol. 6, No. 3, 1959, pp. 423–448.
6. Georgeson, E. H. M., "The Free Streaming of Gases in Sloping Galleries," *Proceedings*, Royal Society London, Vol. A 180, 1942, pp. 484–493.
7. Kersey, D. G., and Hsü, K. J., "Energy Relations of Density-Current Flows: An Experimental Investigation," *Sedimentology*, Vol. 23, 1976, pp. 761–789.
8. Löfquist, K., "Flow and Stress Near an Interface Between Stratified Liquids," *Physics of Fluids*, Vol. 3, No. 2, Mar.–Apr., 1960, pp. 158–175.
9. Middleton, G. V., "Experiments on Density and Turbidity Currents. II Uniform Flow of Density Currents," *Canadian Journal of Earth Sciences*, Vol. 3, 1966, pp. 627–637.
10. Smith, P. C., "A Stream Tube Model for Bottom Boundary Currents in the Ocean," *Deep-Sea Research*, Vol. 22, 1975, pp. 853–873.
11. Wilkinson, D. L., "Studies in Density Stratified Flows," *Report No. 118*, Water Research Laboratory, University of New South Wales, New South Wales, Australia, 1970.
12. Worthington, L. V., "The Norwegian Sea as a Mediterranean Basin," *Deep-Sea Research*, Vol. 17, 1970, pp. 77–84.

APPENDIX II.—NOTATION

The following symbols are used in this paper:

$$A = \text{mass deficit flux;}$$
$$B = \text{width;}$$
$$c_{index} = \text{constant;}$$
$$f = \text{friction factor;}$$
$$f_c = 2\omega \sin\Phi = \text{Coriolis parameter;}$$
$$F_\Delta^2 = \text{densimetric Froude number squared;}$$
$$g = \text{acceleration of gravity;}$$
$$H = \text{total water depth;}$$
$$h = \text{vertical extension of dense bottom current;}$$
$$I_0 = \text{bottom slope;}$$
$$i = \text{index for interface;}$$
$$l = \text{length scale;}$$
$$n = \text{coordinate (perpendicular to flow direction } s);$$
$$p = \text{excess pressure (above ambient fluid pressure);}$$
$$Q = \text{discharge;}$$

$$R_i = F_\Delta^{-2} = \text{Richardson number;}$$

R_f^T = new bulk flux Richardson number;

s = coordinate (perpendicular to n) in flow direction;

V = depth average velocity;

V_E = entrainment velocity;

w = index for wall;

X = horizontal coordinate;

x_1, x_2, x_3 = Cartesian coordinates;

Y = horizontal coordinate;

y = depth;

Z, z = elevation;

β = pressure distribution coefficient, or an angle;

Δ = $(\rho - \rho_0)/\rho_0$ = dimensionless mass deficit;

ρ = density of mass;

σ_t = $\rho - 1000$ (ρ in kilograms per cubic meter);

τ = Reynolds stress;

Φ = latitude;

ϕ = angle (bottom slope); and

ω = rotation of earth.

Prog. Rep. 61, pp. 47-54, Sept. 1984
Inst. Hydrodyn. and Hydraulic Engrg.
Tech. Univ. Denmark

4. LABORATORY EXPERIMENTS ON ENTRAINMENT

DUE TO FREE CONVECTION

by

Fl. Bo Pedersen and Carsten Jürgensen

INTRODUCTION

Free penetrative convection is encountered in many geophysical fields where an air or water volume is exposed to an
unstabilizing heating, cooling, evaporation, freezing etc. The
associated buoyancy flux creates a highly turbulent layer, which
penetrates the non-turbulent ambient fluid of stable or neutral
stratification. A theory for this rate of erosion of the interface - the entrainment velocity V_E - has been presented in
Bo Pedersen [1980], where laboratory and field data reported
on in the literature are given as well.

The present laboratory experiments, which have been performed in a new-built stratified flow flume, see Bo Pedersen
[1984], extend the range in which the theory has been documented
by a decade towards the values often encountered in oceanography.
Furthermore, the accuracy of the present data is higher than
that of the previously reported data in the literature.

THE EXPERIMENTAL SET-UP

The running of the present experiments needs no modifica-
tion of the basic flume set-up. The inlet weir is closed (and
checked for tightness). The outlet overshot weir is raised to a
level slightly higher than the filter in the top-elements, and
then checked for tightness. The tightness of the flume, especially the part which contains the stagnant, ambient salt water,
is crucial, as any leakage from here erroneously would be interpreted as entrainment.

Initially, the flume is filled with salt water up to a
certain level (approximately 10 cm below the filter) and with
tap water up to the filter level. Salt water (Q_W) is then even-

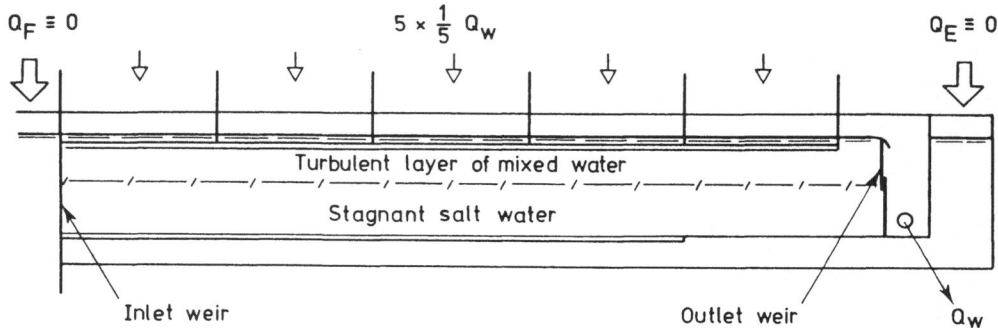

$Q_F \equiv 0$ $5 \times \frac{1}{5} Q_W$ $Q_E \equiv 0$

*Fig. 1 The multipurpose stratified flow flume,
Bo Pedersen [1984], arranged for entrainment measurements in free convection.*

ly poured into the five top-boxes at a constant rate. As no compensating discharge (Q_E) is supplied, a non-stationary process
takes place with a continuous increase in the upper layer salinity (due to the buoyancy flux and the entrainment from below)
and in the depth (due to entrainment solely). The rate of erosion of the interface (equal to the entrainment velocity V_E) can
be measured visually. An independent measure of the upward directed buoyancy flux ($\sim V_E$) can be performed by measuring the
mixed layer salinity as a function of time, which is elucidated
below.

THEORETICAL BACKGROUND

The non-stationary process is described by the conservation equations for

$$\text{mass: } bL \frac{d}{dt} \{ \Delta y + \Delta_E (D-y) \} = (\Delta_W - \Delta) Q_W \tag{1}$$

or by introducing the symbols in Fig. 2

$$- \delta \frac{dy}{dt} - y \frac{d\delta}{dt} = (\delta_E + \delta) v_W \tag{2}$$

and

$$\text{volume: } V_E = \frac{dy}{dt} \tag{3}$$

respectively.

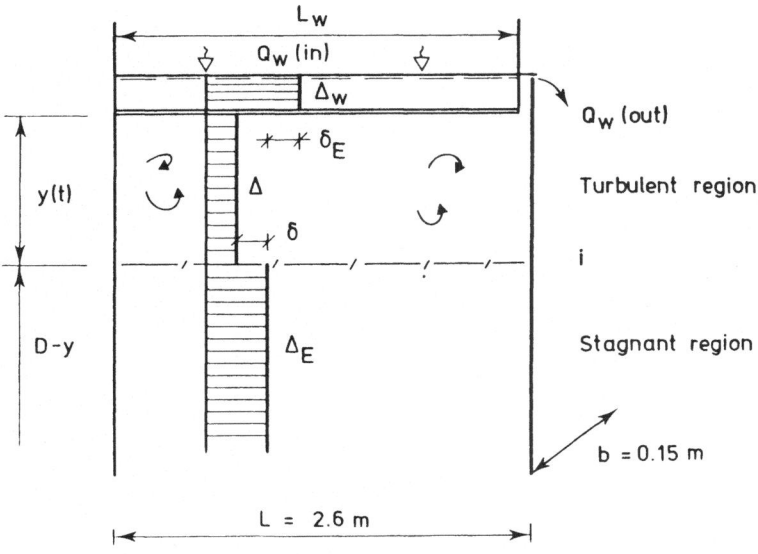

Fig. 2   *Definition sketch (highly distorted scale).*
         Symbols used:
         $\Delta = (\rho - 1000)/1000, \; \rho \, [kg/m^3]$
         $\delta = \Delta_E - \Delta; \; \delta_E = \Delta_W - \Delta_E; \; v_W = Q_W/bl.$

According to the entrainment hypothesis by the senior author (Bo Pedersen [1980]) the buoyancy flux from below $(bl\delta V_E)$ constitutes a certain ratio $(\mathbb{R}_f^T = 0.18)$ of the imposed buoyancy flux from above $((\delta_E + \delta)Q_W)$ in the actual flow range. One of the objectives of the present experiments is to verify (and/or to establish a better estimate of) the efficiency constant $\mathbb{R}_f^T$ in a range, where no measurements have been performed up to date. The entrainment hypothesis states:

$$\delta V_E = \mathbb{R}_f^T (\delta_E + \delta) v_W \tag{4}$$

Combining equations (2), (3) and (4) yields

$$- y \frac{d\delta}{dt} = (\delta_E + \delta) v_W (1 + \mathbb{R}_f^T) \tag{5}$$

and

$$\delta \frac{dy}{dt} = (\delta_E + \delta) v_W \mathbb{R}_f^T = - \left(\frac{\mathbb{R}_f^T}{1 + \mathbb{R}_f^T} \right) y \frac{d\delta}{dt} \tag{6}$$

respectively. Eq. (6) is readily solved to yield

$$\left(\frac{y}{y_0}\right) = \left(\frac{\delta_0}{\delta}\right)^{\frac{\mathbb{R}_f^T}{1+\mathbb{R}_f^T}} \tag{7}$$

(where subscript 0 stands for initial values).

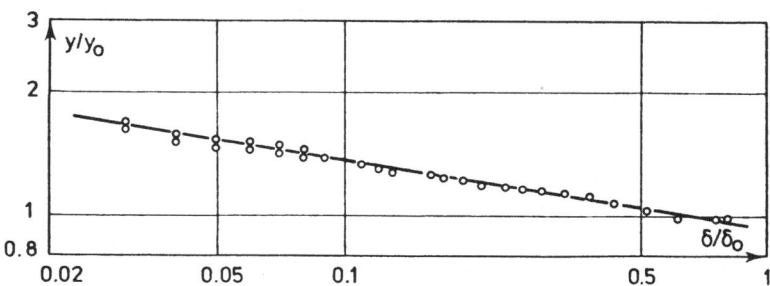

Fig. 3   The y, δ relationship as observed/measured in a typical experiment.

Hence, one way to obtain the efficiency $\mathbb{R}_f^T$ is to plot the (y, δ) observations/measurements in a log-log diagram, see the example Fig. 3.

Another - and partly independent estimate may be obtained by direct measurements of the rate of erosion $V_E = dy/dt$. When using the same salt solution for fluid W and E (i.e. $\delta_E \equiv 0$) we produce a specific simple solution to Eq. (6)

$$y = y_0 + v_W \mathbb{R}_f^T t \tag{8}$$

In Fig. 4 a typical experimental determination of $\mathbb{R}_f^T$ is shown by using Eq. (8).

All the basic experimental parameters are given in table 1. A full report on the experiments is given in Jürgensen [1984].

In order to compare the present data with the data available in the literature, they have both been plotted in Fig. 5.

In the actual range ($F_{\Delta,W}^2 \ll 1$) a constant $\mathbb{R}_f^T$ corresponds to a straight line (slope 1:1) in Fig. 5, which is easily

Exp.	Δ_E 10^{-3}	Y_0 m	V_W 10^{-5}m/s	V_E 10^{-5}m/s	time start stop min.	δ 10^{-3}	$\frac{Y}{Y_0}$	W_F 10^{-3}m/s	$\mathbb{F}_{\Delta,W}^2$ 10^{-3}	V_E/W_F 10^{-3}
a	37.7	0.117	48.3	9.1	2 / 12	17.7 / 2.26	1.06 / 1.51	26.4 / 15.0	22.2 / 39.1	3.44 / 6.08
b	37.8	0.007	6.82	1.39	10 / 45	18.1 / 3.40	1.10 / 1.48	10.1 / 6.41	6.73 / 10.7	1.37 / 2.17
c	37.1	0.090	3.71	0.574	20 / 175	16.0 / 0.742	1.08 / 1.67	8.20 / 3.42	4.53 / 10.8	0.70 / 1.68
d	37.4	0.100	21.3	3.33	2 / 14	28.1 / 5.98	1.04 / 1.27	18.3 / 11.7	11.7 / 18.3	1.82 / 2.85
e	35.8	0.100	60.5	11.0	4 / 13	15.8 / 1.07	1.24 / 1.84	22.6 / 10.6	26.7 / 57.4	4.86 / 10.4
f	35.6	0.100	11.1	1.56	5 / 25	25.6 / 6.05	1.18 / 1.24	14.3 / 9.40	7.73 / 11.8	1.09 / 1.66

Table 1. Tabulated values of experimental conditions and findings in the experiments by Jürgensen [1984]. The values correspond to the initial and final stage

$$\Delta_E = \frac{\rho_E - \rho_R}{\rho_R} \quad ; \quad \rho_R = 10^3 \text{ [kg/m}^3] \quad ; \quad W_F = \left(gy\delta V_w \right)^{1/3}$$

$$\delta = \frac{\rho_E - \rho}{\rho_R} = \Delta_E - \Delta \quad ; \quad \mathbb{F}_{\Delta,W}^2 = \frac{W_F^2}{gy\delta}$$

$$V_E = \frac{dy}{dt}$$

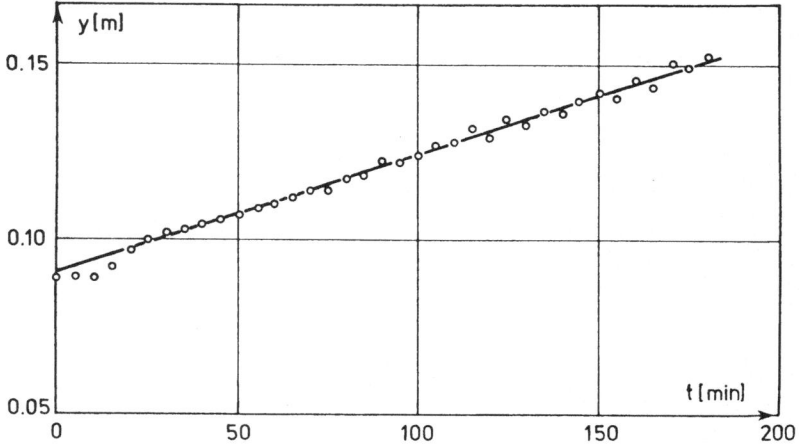

*Fig. 4   The observed erosion of the interface in a
           typical experiment.*

verified by using the theoretical expression   (see Fig. 5)

$$\frac{V_{E,L_W}}{W_F} \simeq I\!R_f^T \; I\!F_{\Delta,W}^2 \quad (I\!F_{\Delta,W}^2 \ll 1) \tag{9}$$

which yields

$$V_{E,L_W} = I\!R_f^T \; \frac{W_F^3}{gy\delta} = I\!R_f^T \; \frac{gy(\delta_E+\delta)}{gy\delta \; bL_W} \; Q_W \tag{10}$$

or simply

$$\delta V_{E,L_W} = I\!R_f^T \; (\delta_E + \delta) v_W \; \frac{L}{L_W} \tag{11}$$

where V_{E,L_W} is the entrainment velocity on the length L_W. Due to
the experimental set-up, entrainment takes place over a slightly
greater length L, which means that the measured rate of erosion
has to be reduced by the factor L_W/L, and hence Eq. (11) and (4)
are identical.

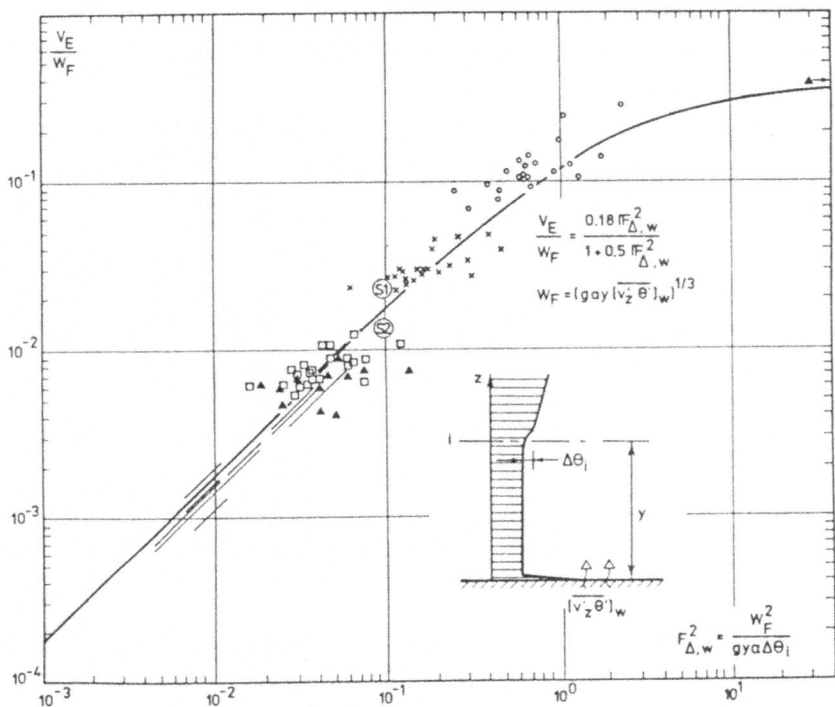

Fig. 5 *Laboratory and field data on entrainment for free*
penetrative convection compared with theory
(Bo Pedersen, 1980).

The points are based on data referred by:

o, x, □   *Heidt [1975] (laboratory experiments).*

S1, S2   *Willis and Deardorff [1974] (laboratory*
experiments).

▲   *Farmer [1975] (Field data from solar*
heating beneath lake ice).

/   *Present laboratory experiments.*

Comments: $\alpha\Delta\theta_i = \delta$ *in the present experiments.*

$\alpha[\overline{v'_z\,\theta'}]_W = (\delta_E + \delta)v_W$ *in the present experiments.*

COMMENTS TO THE RESULTS

An efficiency factor of $\mathbb{R}_f^T = 0.153 \pm 0.02$ was evaluated, which is slightly less than the value suggested previously. A similar value (0.150) has been found for jets by Sehested [1982] and by the senior author (0.155, for jets too). Hence, there seems to be some experimental evidence for using $\mathbb{R}_f^T = 0.15 - 0.16$ instead of 0.18 in the future. The physical implication of this correction is insignificant - contrary to the physical implications of the major findings in the present experiments, namely that the bulk flux Richardson number theory applies in the very low range of entrainment encountered in nature.

REFERENCES

Bo Pedersen, Fl. [1980]. "A monograph on turbulent entrainment and friction in two-layer stratified flow". Series Paper No. 25, Inst. of Hydrodynamics and Hydraulic Engineering, Tech. Univ. of Denmark.

Bo Pedersen, Fl. [1984]. "A multipurpose stratified flow flume". Progress Rep. No. 61, Inst. of Hydrodynamics and Hydraulic Engineering, Tech. Univ. of Denmark.

Jürgensen, C. [1984]. "Experimental investigation on mixing processes in two-layer stratified flow". Master Thesis (in Danish). Supervisor Fl. Bo. Pedersen.

Sehested, J. 1982 . "A note on entrainment in surface jets discharging into a flowing recipient and exposed to the wind". (in Danish). Danish Hydraulic Institute, DK-2970 Hørsholm, Denmark.

Prog. Rep. 28, pp. 31-38, April 1973
Inst. Hydrodyn. and Hydraulic Engrg.
Tech. Univ. Denmark

6. STEADY WIND SET-UP IN PRISMATIC LAKES

by Frank Engelund

ABSTRACT

The paper discusses the velocity distribution in a channel when a shear stress acts at the water surface. The result is applied to account for the secondary currents in a lake of prismatic shape when the water is assumed to be homogeneous.

VELOCITY DISTRIBUTION IN PLANE COUETTE FLOW

We consider the case of a steady and plane Couette flow induced by the motion of the upper plate, see Fig. 1.

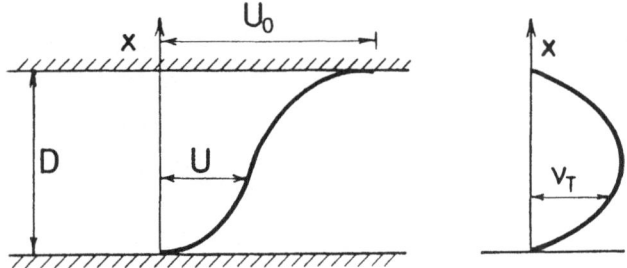

Fig. 1. Velocity profile and assumed distribution of eddy viscosity

The distance between the plates is D, the velocity of the moving plate is U_0. The shear stress is constant and equal to τ_0. Assuming applicability of the eddy viscosity concept for the considered flow, we get

$$\tau_0/\rho = U_f^2 = \nu_T \frac{dU}{dx} \tag{1}$$

in which U_f is the friction velocity. A plausible distribution of ν_T has been suggested by H. Reichardt [1]:

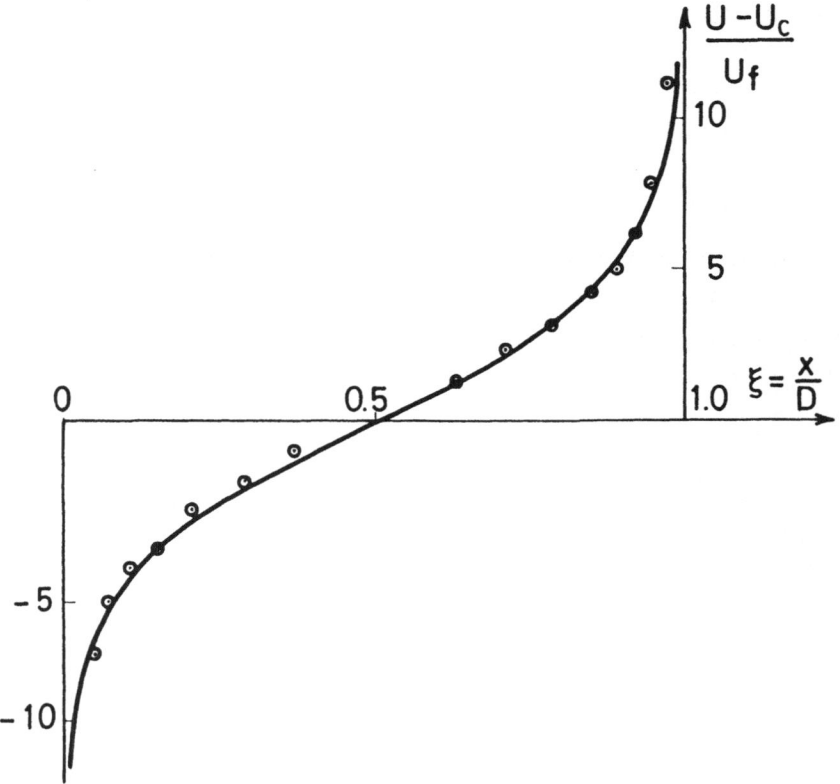

Fig. 2   Comparison between theory and measure-
          ments by H. Reichardt, [1]

$$\nu_T = \kappa U_f \, x \, (1 - \frac{x}{D}) , \tag{2}$$

which is the simplest distribution that has the correct value near the boundaries. By substitution in Eq. 1 we derive the following expression

$$\frac{U}{U_f} = \frac{1}{\kappa}[\ln \xi - \ln(1 - \xi)] + c \tag{3}$$

in which c is an arbitrary constant and $\xi = x/D$.

Near the lower plate the velocity distribution is assumed to approach the classical logarithmic distribution asymptotically, which yields the following formula

$$\frac{U}{U_f} = \frac{1}{\kappa} \ln \frac{\xi}{1-\xi} + \frac{1}{\kappa} \ln \frac{30D}{k} , \tag{4}$$

where k is Nikuradse's equivalent roughness for the lower plate. Hence, at the distance $k/30$ from the theoretical plate surface we get zero velocity.

To obtain an expression for velocity U_0 of the upper plate we put $x = D - k_0/30$, where k_0 is the roughness of upper plate:

$$\frac{U_0}{U_f} = \frac{1}{\kappa}[\ln \frac{30D}{k_0} + \ln \frac{30D}{k}] \tag{5}$$

If the plates are equally rough, the profile is symmetric about the centre.

The theory is compared with experiments [1] in Fig. 2. U_c denotes the velocity in the middle of the channel, where $\xi = \frac{1}{2}$.

CHANNEL FLOW WITH WIND STRESS ALONG THE WATER SURFACE

We consider a steady two-dimensional flow in a channel assuming that a shear stress τ_s is acting in the flow direction ($\tau_s > 0$) - or against the flow direction, $\tau_s > 0$. The velocity distribution is supposed to be given by an expression of the form

$$U = a \ln \xi - b \ln (1 - \xi) + c \tag{6}$$

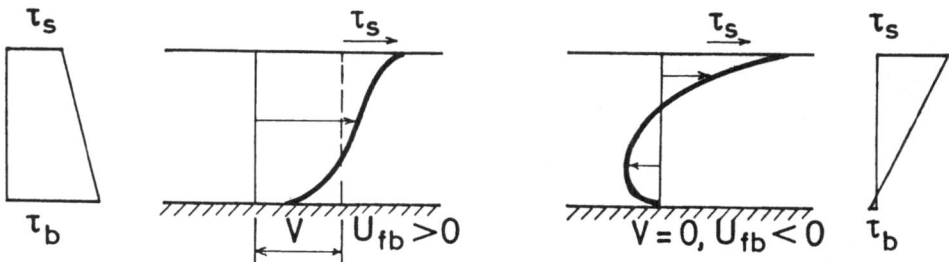

*Fig. 3. Shear stress distribution and velocity distri-
bution in a channel and a lake, respectively.*

The following items support the plausibility of this
expression:

1) As before, we consider the flow near the bottom
   and Eq. 6 is seen to approach the classical logarithmic velocity profile if we take

$$c = a \ln \frac{30D}{k} \quad \text{and} \quad a = \frac{U_{fb}}{\kappa} \qquad (7)$$

   U_{fb} being the shear velocity at the bottom, positive
   if the flow near the bed is downstream.

2) The formula gives the logarithmic distribution near
   the water surface if we take

$$b = \frac{U_{fs}}{\kappa} \qquad (8)$$

   where U_{fs} is the shear velocity at the surface (positive if $\tau_s > 0$).

3) The equation reduces to that of ordinary channel
   flow if $\tau_s = 0$.

4) If $\tau_s = \tau_b$, we get the distribution for plane Couette
   flow. Substitution of Eqs. 7 and 8 gives

$$U = \frac{U_{fb}}{\kappa} \ln \frac{30x}{k} - \frac{U_{fs}}{\kappa} \ln (1 - \xi) \qquad (9)$$

From this we find that the surface velocity becomes

$$U_0 = \frac{U_{fb}}{\kappa} \ln \frac{30D}{k} + \frac{U_{fs}}{\kappa} \ln \frac{30D}{k_0} \qquad (10)$$

where k_0 is the surface roughness (about 50 mm).

By integration of Eq. 9 we find for the mean velocity V over the depth ($\kappa = 0.40$)

$$V = 6.0\, U_{fb} + 2.5\, U_{fb} \ln \frac{D}{k} + 2.5\, U_{fs} \qquad (11)$$

from which we get the important expression

$$U_{fb} = \sqrt{\frac{f}{2}}(V - 2.5\, U_{fs}), \qquad (12)$$

where f is the friction factor for the bottom, given by

$$\sqrt{\frac{2}{f}} = 6.0 + 2.5 \ln \frac{D}{k}. \qquad (13)$$

A similar expression has been suggested by Ian Larsen [2].

An interesting special case is the steady wind set-up in a rectangular lake for which case we have V = 0 and, consequently, from Eq. 12 we get

$$\frac{U_{fb}}{U_{fs}} = -2.5 \sqrt{\frac{f}{2}} \quad \text{or} \quad \left|\frac{\tau_b}{\tau_s}\right| = 3.25\, f \qquad (14)$$

This shows that the bed shear stress is of an order of magnitude of about 1 or 2 per cent, in agreement with measurements [3].

CIRCULATION IN A PRISMATIC LAKE

For a long prismatic lake and a longitudinal wind set-up (see Fig. 4) it is possible to obtain some very simple results concerning the circulation pattern. We apply a coordinate system with x_1-axis along the axis of the lake, x_2 horizontal and perpendicular to it and x_3 vertical in downward direction. The local depth is y (as indicated in the cross section, Fig. 4 left, the local level of the bed is z. The flow equations then read

$$\frac{\partial}{\partial x_1}\int_0^y v_1^2\, dx_3 + \frac{\partial}{\partial x_2}\int_0^y v_1 v_2\, dx_3 = -gy\frac{\partial}{\partial x_1}(z+y) + \frac{\tau_s}{\rho} - \frac{\tau_{b_1}}{\rho} \qquad (15)$$

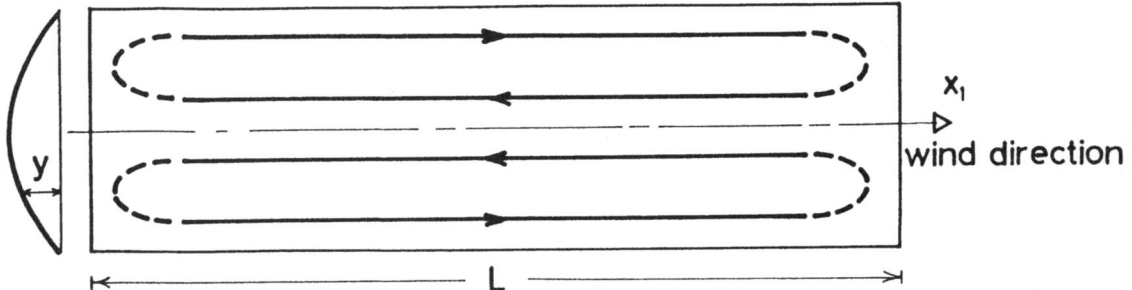

Fig. 4. Main circulation pattern in prismatic lake
(schematically)

$$\frac{\partial}{\partial x_2}\int_0^Y v_2^2 dx_3 + \frac{\partial}{\partial x_1}\int_0^Y v_1 v_2 dx_3 = -gy\frac{\partial}{\partial x_2}(z+y) - \frac{\tau_{b_2}}{\rho} \qquad (16)$$

$$\frac{\partial}{\partial x_1}\int_0^Y v_1 dx_3 + \frac{\partial}{\partial x_2}\int_0^Y v_2 dx_3 = 0 \qquad (17)$$

If the lake is sufficiently long (in the wind direction), a region of uniform flow will occur in the middle
part and for this region a simple solution of the equation
is given below. Near the ends of the lake, on the other hand,
the flow pattern is extremely complicated and will not be
considered here.

In case the lake has a broad rectangular cross section,
the wind set-up s is given by the well-known expression

$$\frac{s}{L} = \frac{\partial}{\partial x_1}(z+y) = \frac{\partial s}{\partial x_1} = \frac{\tau_s - \tau_b}{\rho g D} \qquad (18)$$

where L is the length and D the depth of the lake. τ_b is obtained from Eq. 14. This formula indicates a wind set-up inversely proportional to the depth. In case the depth is varying over the cross section (as in Fig. 4), this formula -
if applied uncritically - indicates a non-uniform distribution of the set-up, increasing from a relatively small value
in the middle of the lake to infinity at the beaches. This
is, of course, a picture very far from reality. What actual-

ly occurs is a uniform set-up with $\partial s/\partial x_1 = s'$ constant over the cross section and a set of secondary currents, as explained below.

For the central part with uniform flow, the terms on the left-hand side of Eqs. 15 - 17 vanish. The only remaining equation is

$$gy\frac{\partial s}{\partial x_1} = g\,y\,s' = \frac{\tau_s - \tau_b}{\rho}$$

from which we find that

$$\frac{\tau_b}{\tau_s} = \frac{\tau_s - \gamma\,y\,s'}{\tau_s} \tag{19}$$

Now, we introduce the substitution $y = D + \eta$ and define the depth D so that the following relation is fulfilled:

$$\tau_s = \gamma\,D\,s' \tag{20}$$

After insertion in Eq. 19 this reduces to

$$\frac{\tau_b}{\tau_s} = -\frac{\gamma\eta}{\tau_s}\,s'$$

or by extracting the square root

$$\frac{U_{fb}}{U_{fs}} = \text{sign}(\eta)\sqrt{|\frac{\eta}{D}|} \tag{21}$$

When this is combined with Eq. 12, we are able to find the local mean velocity V in any vertical and the last problem is to evaluate the depth D. This is done by expressing the condition that in the case of steady flow there is no net transport of water through a cross section

$$\int_0^B V(D + \eta)\,dx_2 = 0 \tag{22}$$

As an illustrative example a lake with parabolic cross section and $\sqrt{2/f} = 20$ was chosen. Some results of the calculations are given in Fig. 5.

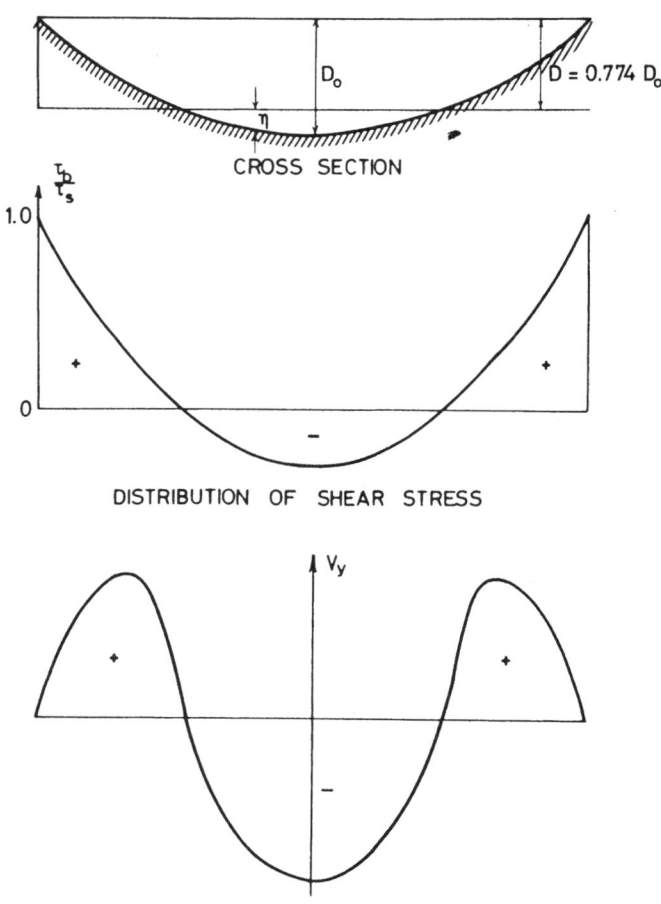

CROSS SECTION

DISTRIBUTION OF SHEAR STRESS

DISTRIBUTION OF VOLUME FLUX (PER UNIT WIDTH)

Fig. 5 Numerical example

References:

[1] Reichardt, H.: Gesetzmässigkeiten der geradlinigen tur-
    bulenten Couetteströmung. Mitteilungen aus dem Max-
    Planck-Institut für Strömungsforschung und der Aero-
    dynamischen Versuchsanstalt. No. 22, Göttingen 1959,
    pp. 1 - 45.

[2] Larsen, Ian: Om Tolagsstrømninger I. Vinds effekt på
    grunde vandområder. (Thesis in Danish). Coastal En-
    gineering Laboratory. Technical University of Denmark,
    1962.

[3] Francis, J.R.D.: A note on the velocity distribution and
    bottom stress in a wind-driven water current system.
    Sears Foundation: Journal of Marine Research, Vol. 12,
    No. 1, May 15, 1953, pp. 93-98.

Nordic Hydrology, 12, 1981, 1-20

Diversion of the River Neva

How will it Influence the Baltic Sea, the Belts and Cattegat

Flemming Bo Pedersen and Jacob Steen Møller

Technical University of Denmark, Copenhagen

Diverging part of the river Neva discharge to the dry regions in the southern USSR has raised the question, to what extent such a river diversion will influence the hydrographic conditions in the Baltic Sea and the Danish Inland waters. In order to quantify the influence, the system has been divided into eight subareas, each of which is characterized by an equation for the mass, the volume and the dynamic balance (the mixing), respectively. The man-made change in the river runoff has been introduced in the equations, which have then been linearized and solved with respect to changes in the salinities, the discharges and the layer depths in the system.

As a quantitative example the hydrographic consequences of a 25% reduction in the river Neva discharge have been outlined. The most pronounced influence is on the salinities, which are increased by 0.2 to 0.4 ‰ allover in the system. Hence, if the river diversion had become executed in the beginning of this century a 30 to 40% higher salinity-variation would have been encountered in the Baltic Sea – compared to the actual variations during this century.

Introduction

The increasing water demand for irrigational purposes in the dry regions north of the Caspian Sea and the Lake Aral (in the USSR) has actualized the plans of pumping huge amounts of water from the catchments of the river Ob and the river Neva to the river Volga, which are running through the affected dry areas. The USSR's Council of Ministers have, in fact, in their 5-year plan 1976-80 initiated the preliminary planning for diverging up to 2,000 m³/s from the river Ob, which drains to the Arctic Sea. Although not mentioned directly in the available sparse information on the project, Mikhaylow et al. (1977), Voropaev (1978), Golubev (1978), it is obvious from an engineering point of view, that the river Neva is also attractive as a source to this irrigation project. With a discharge of approximately 3×10^3 m³/s, the river Neva is the largest single fresh water contributor to the Baltic Sea, to which the total average fresh water input is, in the order of 15×10^3 m³/s. Therefore, a radical decrease in the runoff from the river Neva has a great bearing on the hydrography of the Baltic. Further, it has also a large effect on the hydrography of the inland Danish waters, which links the brackish Baltic Sea to the ocean.

A man-made regulation of the river Neva is therefore a matter of international concern, as it will influence all the countries boardering the Baltic. On the other hand, there seems to be no international laws or conventions, which makes it possible for the other Baltic countries to change the decisions if possibly unwanted effects of the regulations can be foreseen. The problem has a parallel in the Danish project for building a bridge across the Great Belt, which was estimated to have a measureable influence on the Baltic Sea, Bo Pedersen (1978). Although there was an international reaction against the building of the bridge, it was for economic reasons, that the Danish government finally decided to postpone the bridge project.

The main objective of the present paper is to establish an estimate of the hydrographic changes in the Baltic Sea and the Danish inland waters if part of the river Neva's discharge is diverged from the Baltic Sea. An evaluation or estimation of the possible consequences for the affected countries is beyond the scope, but it is the hope, that the article will act as a trigger for further discussions, and that the findings will serve as a basis for further work.

The Basic Principles and Assumptions for the Model

In the Baltic Sea and its connections with the North Sea (the Cattegat, the Belts and the Sound) all types of estuaries, i.e. semienclosed bodies of water, where a measurable dilution by fresh water are present, can be recognized. Although a throughout hydrographic description of an estuary demands knowledge of the variation in space and time of all relevant physical properties, such as salinity, temperature, oxygen content, phosphate and nitrate concentrations etc. we shall make a common approach and restrict ourselves to a representative *steady-state* situation considering only the *salinity* distribution, which is the property governing the vertical stability and hence the mixing in the actual case. The most simple representation of an estuary in which the basic physical conditions are maintained is a *two-layer* flow. An inspection of the actual conditions in the Baltic Sea and in the Danish inland waters confirms, that this is a fair approximation.

Our approach is then, first to identify the major external forces affecting the system (fresh water discharge, wind, tide, etc.), then to estimate the correct order of magnitude of the strength of these forces, introduce them in our model and then finally confirm with the actual measured conditions in the estuary, that our model is reasonable representative for the dynamics of the estuary. After the verification of the model, we introduce the change in the fresh water discharge from the river Neva – linearize the equations – and solve with respect to the changes in the salinities, the depths and the flows in the idealized estuaries. In these calculations we have focused on the man-made changes in the fresh water discharge. The consequences for the layer depths and salinities in the Baltic Sea for a natural variation in the fresh water discharge are different from our findings, due to the strong correlation between the precipitation (and hence the runoff) and the meteorological conditions, the last being held unchanged in our calculations.

As stated above all types of estuaries are present in the model. The dynamics of an estuary is mainly affected by the following parameters, Bo Pedersen (1980a)

1. The geometry
2. The hydrology of the adjacent watershed
3. The oceanographic conditions outside the estuary
4. The wind field (and the barometric pressure variation due to the large dimensions of the Baltic Sea).

The great variability of these parameters over the actual oceanographic field makes it necessary to divide the total area into eight subareas as indicated in Fig. 1. The subdivision is chosen in such a way that a reasonable simple dynamic description can be given for each region, and hence, the areas do not represent regions of equal importance, merely areas of different dynamic behaviour.

For each subarea steady-state continuity-equations for mass and volume are established. One of the terms of major importance for the continuity equations, is the term representing the mixing accross the interface separating the two layers. This mixing is due to the generation of turbulence by external forces, such as tide, variable meteorological conditions, etc., i.e. all highly non-stationary forces. Therefore, although the basic objective is to establish a steady-state model, it is necessary to incorporate the non-steady dynamic behaviour of the system in the description in order to maintain the correct physics. To transfer a dynamic situation to an artificial steady state demands knowledge of the representative time scale and the representative force scale. With focus on the mixing, a representative averaging time scale is the residence time, T, i.e. a measure of the mean time that a particle of tracer remains inside the actual subarea of the estuary system

$$T = \frac{Vol}{Q} \qquad (1)$$

where Vol is the total volume of pure fresh water inside the subarea and Q is the accumulated fresh water discharge at the actual cross section. The residence time for the Baltic estuary system varies from for example, typically a week in the Belt

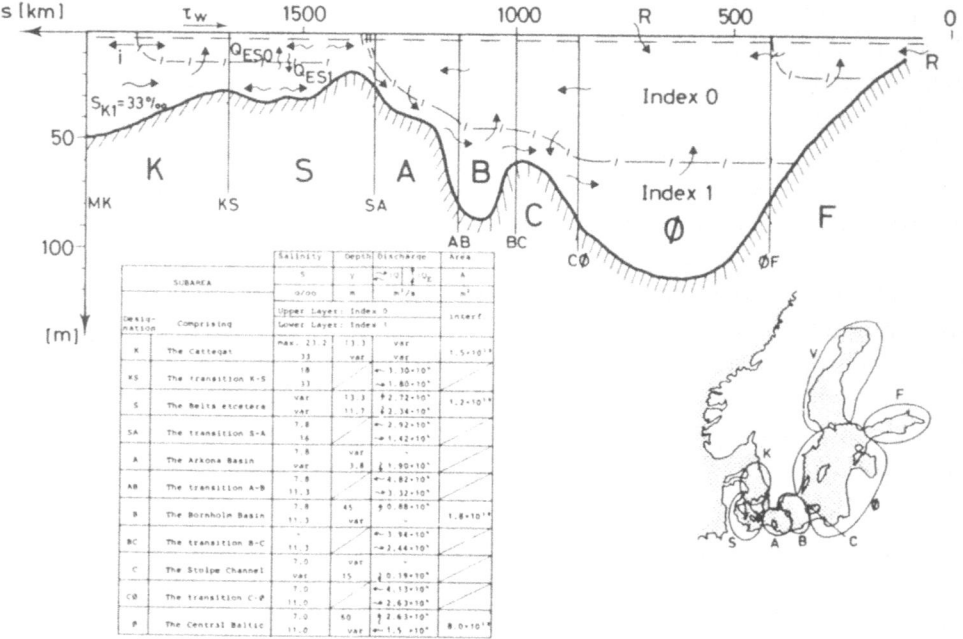

Fig. 1. The Baltic estuary system divided into eight subareas. The specific hydrodynamic characteristics of the six outermost subareas are summarized in the table.

region, a month in the Cattegat region to 30 years at the central Baltic. As the time scale for the tide (a day) as well as for an average meteorological event (a week) are below the averaging time for the estuary, these two types of external forces can in the time frame be treated as steady, persistent forces, although it may be admitted, that the seasonal variations as for instance in the meteorological activity and in the runoff cannot be incorporated in our theory. On the other hand, the seasonal variations are much weaker than the single events, and can therefore be neglected in the analysis. The other important scale for the mixing is the force scale, i.e. a measure for the energy available for the mixing process. This is the subject of the next chapter.

Mixing in a Two-Layer Stratified Flow

The two-layer stratified flow is characterized by having two nearly homogeneous layers separated by an interface with a sharp density gradient. The mixing between the two layers can be treated as pure (one-way) entrainment if the level of kinetic energy is high in the one layer and negligible in the other layer. If a measurable level of kinetic energy is present in both layers a two-way transport exists, which can be treated either as a combined entrainment/diffusion problem or, as we prefer it, as a double-sided entrainment. A comprehensive analysis of the entrainment functions for a large class of two-layer flows can be found in Bo Pedersen (1980a). The basic assumption for all the flow cases treated there is, that a universal relationship exists between the energy available for the turbulence (i.e. the production with some minor corrections) and the energy gained (potential as well as turbulent kinetic energy) due to the entrained mass. Hence, the characteristic force scale for the mixing, i.e. for the entrainment, can be evaluated by taking a moving average value of the energy input into the system, which by Bo Pedersen (1980a) is shown to be proportional to the mean speed in the layer $|v|$ to the third power. Hence, the proper dynamic transformation from the non-steady to the steady system is done by applying a mean velocity $\tilde{V}$ defined by

$$\tilde{v} = \left(\frac{1}{T} \int_0^T |v|^3 dt \right)^{\frac{1}{3}} \tag{2}$$

The velocity scale in the continuity equation is the simple mean velocity and not the velocity defined by Eq. (2). It is therefore necessary to incorporate a circulation-velocity with no net transport inside some of the regions in order to get dynamic- as well as mass-balance in the simplified systems.

The major external forces producing turbulence in the system are:

1. *The wind,* which generates a flow in the upper layer. A persistent wind acting far from boundaries causes an entrainment velocity V_E which can be evaluated by the following equation, Bo Pedersen (1980a)

$$\frac{V_E}{U_F} = \frac{2.3}{6 + \mathbb{R}i_F} \qquad \mathbb{R}i_F = \frac{\Delta g y}{U_F^2} \tag{3a}$$

where $U_F = (\tau_w/\varrho)^{1/2}$ is the friction velocity in the water due to the windstress τ_w. The bulk Richardson number $\mathbb{R}i_F$ is a measure of the stability of the system as Δ is the non-dimensional density difference between the upper and the lower layer ($\Delta\varrho = \varrho_{lower} - \varrho_{upper}$), g is the acceleration of gravity and y the upper layer depth. All the subareas in the Baltic have rather stable interfaces, i.e. $\mathbb{R}i_F >> 6$, which means that Eq. (3) can be reduced to

$$\frac{V_E}{U_F} \simeq \frac{2.3 \, U_F^2}{\Delta g y} \tag{3b}$$

2. *The heating/cooling* process forms during the summertime a stable thermocline. In the winter period it creates an unstable free convection, which erodes the thermo- or halocline.

As shown by Bo Pedersen (1980a) it is only in those parts of the Baltic system, where the halocline is located deep (Bornholm Basin $y \simeq 45$ m, Baltic Proper $y \simeq 60$ m), that a thermocline forms during a pronounced period of the year. The thermocline acts as a lid, which prevents the wind from creating mixing through the halocline – in the actual region during nearly half a year, which has to be taken into account in the dynamical part of the calculations. During the thermocline-free period the free convection plays the minor role in the overall erosion of the halocline. Therefore the only influence from the heating/cooling in our simple model is, that it prevents mixing in the Bornholm Basin and in the Baltic Proper during half a year.

3. *The tide* generates a periodic in and out flow, which can be registrated in the Danish inland waters. On the other hand, the energy input into the system from the tide is sufficiently small to be negligible in the present analysis.

4. *The meteorological activities* over Scandinavia with succeeding low and high pressure acts like a piston on the Baltic Sea. Combined with wind set-up and set-down an oscillating in- and out-flow through the Danish inland waters is generated. In the Cattegat, the Belts and the Sound this means that a large part of the surface and the bottom water is pendling in an out producing turbulent kinetic energy and therefore mixing. The other type of mixing, which shall be considered, occurs in the Arkona region where the pendling only takes place in the surface water. The saline bottom water is trapped in a dense bottom current on the eastern slope of the Darss Sill (16 m depth) in the Great Belt and on the southern slope of the Drogden Sill (8 m depth) in the Sound.

The order of magnitude of the non-steady flow in the Cattegat and the Belts can be evaluated from the discharge measurements performed in the Great Belt, reported by Jacobsen (1980), see Fig. 2. The typical amplitude in the pendling is about $10^5 \, m^3/s$, which is 10 times the average fresh water outflow through the Great Belt. This ratio between the mass average and the dynamic average velocity demonstrates the presence of a large no net flow circulation.

The circulation induced mixing can be treated as a quasi-steady mixing due to the extreme large ratio between the non-steady period of the circulation (weeks) and the mixing time scale (hours). For a steady-state condition the strength of the circulation induced entrainment to the wind induced entrainment can be shown (Bo Pedersen 1980a) to be equal for a ratio of the dynamic mean velocity $\tilde{V}$ to the wind generated friction velocity U_F of

$$\frac{\tilde{V}}{U_F} \simeq 50 \tag{4}$$

In the Cattegat a typical high front speed is $\dot{V} \simeq 0.1 \, m/s$, while the representative dynamic friction velocity due to the wind is $U_F \simeq 8 \times 10^{-3} m/s$. Hence in Cattegat the circulation contribution to the mixing is only a few per cent of the wind generated mixing and can therefore be neglected.

In the Belts the typical observed velocities are of an order of magnitude which makes them just as important for the mixing process as the wind, i.e. $V \simeq 0.4 \, m/s$.

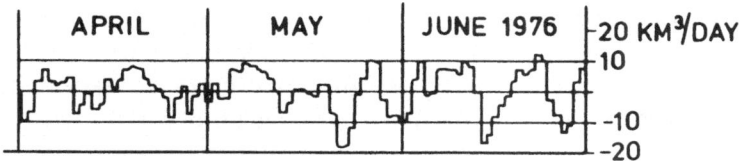

Fig. 2. Typical time series of the measured outwards (positive) and inwards
discharge through the Great Belt. From Bo Pedersen (1978).

Fortunately for the present analysis the circulation as well as the wind do both
originate from the meteorological activity over Scandinavia, which is kept unchanged in the analysis. The above-mentioned theory considering the ratio between
the gain in energy due to entrainment to the production of turbulent kinetic
energy simply states for the Belts, that the volume of entrained water Q_{EO}
amounts to

$$Q_{E0} = \frac{A \cdot constant}{(S_{S_1} - S_{S_0})y_{S_0}} \tag{5}$$

The constant in the numerator stands for the dynamic turbulence production and
is estimated below. The denominator represents the gain in potential energy of
the entrained mass, namely proportional to the salinity difference (the paranthe-
sis) and the upper layer depth. The high velocities in the non-stationary flow in the
Belts creates a downwards as well as an upwards directed entrainment. Again –
using the efficiency concept for the mixing – the downwards entrainment is similar
to the upwards entrainment discharge

$$Q_{E1} = \frac{A \cdot constant}{(S_{S_1} - S_{S_0})y_{S_1}} \tag{6}$$

where the constant stands for the dynamic energy input and y_{S1} is the lower layer
depth in the Belts.

The dense bottom current in the Arkona Basin is a highly intermittent flow (Bo
Pedersen 1977, Petrén and Walin 1975), which only takes place in connection with
an inflow situation to the Baltic. The other dense bottom current in the system –
from the Bornholm Basin through the Stolpe Channel into the Baltic proper – is a
nearly persistent flow, due to the reservoir effect of the Bornholm Basin (Bo
Pedersen 1977, Rydberg 1976). Dense bottom currents in a rotating coordinate
system has been treated in Bo Pedersen (1980b). The discharge Q as a function of
the distance s along the pathline of the flow is increasing due to entrainment, such
that

$$Q(s) = Q(s=0) \exp\{0.072 \int_0^s \frac{I \, ds}{y} \} \tag{7}$$

where I is the local bottom slope. As the depth y is nearly independent of the distance, the integral simply is the drop in elevation non-dimensionalized by the depth. For an unchanged geometry it can furthermore be expected, that the densimetric Froude's number at the sill is unchanged too, see Bo Pedersen (1980a and b), i.e.

$$(\frac{Q^2}{\Delta g B^2 y^3})_{before} = (\frac{Q^2}{\Delta g B^2 y^3})_{after} \tag{8}$$

where the indexes before/after relates to the change of the river Neva discharge into the Baltic Sea.

Characteristics of the Subareas

With the major external forces identified and the associated mixing processes quantified, a description of the individual subareas can be given, including the equations governing the mass and the volume balances and the dynamic behavior. Furthermore, we shall try to a certain extent to verify the simple models presented, against field measurements. To facilitate these assessments a summarizing chart is given, Fig. 1, which contains the symbols used and the values adapted in the present approach to the problem. Finally a summary scheme is given, which contains all the equations needed for the final calculation. The initial values attributed the salinities, the depths and the discharges (see Fig. 1) all satisfy the outlined equations, and are therefore consistent with our interpretation of the dynamics of the system.

Subarea 1: The Cattegat

The western part of the approximately 100 km wide and 240 km long Cattegat has an average depth of approximately 10 m, which is less than the nearly constant upper layer depth in the permanent salt water wedge present in the eastern part, where the total depth is up to 100 m. Hence in the shallow western part a well mixed estuary is normally present, while the eastern part has the characteristics of a two-layer salt water wedge, with constant salinity in the lower layer and varying upper layer salinity.

According to our model, the mixing in the Cattegat is primarily upwards directed entrainment caused by the energy input from the wind. With reference to Fig. 1 the continuity equation for volume states

$$Q_{K0} = Q_{KS0} + \int_0^x V_{EK} B_K \, dx \tag{9}$$

where Q_{K0} is the upper layer net discharge in position x, and Q_{KS0} is the southern boundary value of this discharge (B_K = the interfacial width). The mass deficit flux is constant, because the lower layer salinity is constant and because the mixing is pure upwards directed entrainment. Hence

$$Q_{K_0} \Delta_K = Q_{KS_0} \Delta_{KS} \qquad (10)$$

where Δ_K and Δ_{KS} stands for the non-dimensional mass deficit at position x and at the southern boundary, respectively.

The dynamic conditions in the Cattegat is described by an entrainment function, Eq. (3b) as well as a boundary condition at the northern boundary, where a front is present. In lack of detailed knowledge of the dynamics of this front a common, simple front condition has been used, namely that the densimetric Froude number $I\!F_\Delta$ at the front is a constant, i.e.

$$I\!F_\Delta^2 = \frac{Q^2}{\Delta g B^2 y^3} = constant \qquad (11)$$

where the discharge Q stems from the oscillatory flow.

Eqs. (9), (10) and (3b) can be solved to yield the density-deficit distribution in the Cattegat

$$\frac{1}{\Delta_K} = \frac{1}{\Delta_{KS}} \exp(\frac{x}{\lambda_K}) \qquad (12)$$

where the length scale λ_K is determined by

$$\lambda_K = \frac{\Delta_{KS} \, g \, y_{K_0} Q_{KS_0}}{2.3 \, B_K U_F^3} \qquad (13)$$

If we introduce the empirical relation between the deficits in the density ϱ and the salinities

$$\rho_1 - \rho_0 \simeq 0.75 (S_1 - S_0) \qquad (14)$$

Eq. (1) can be transformed to

$$S_{K_1} - S_{K_0} = (S_{K_1} - S_{KS_0}) \exp (\frac{1.680 \cdot 10^3 - s}{\lambda_K}) \qquad (15)$$

where s (in m) is the overall stationing which starts at the head of the Bothnian Bay (at the Neva inlet to the Bay) and end at the Cattegat/Skagerack front where s = 1,920 × 10³ m (s = 1,680 × 10³ m corresponds to the Belt/Cattegat transition), and λ_K takes the value 565 × 10³ m.

Eq. (15) describes an equivalent steady, yearly averaged salinity-distribution in the upper layer of the Cattegat. This layer is in fact subject to great forwards and backwards movements during the year. Hence Eq. (15) cannot be checked by field measurements before it has been modified slightly. The non-steady movements are reflected in the position of the Cattegat front, which during an inflow situation moves towards the Belts. North to the front the saline Skagerack water is encountered (salinity ≈ 33‰) – south to the front the brackish Cattegat water is present. The front movements have been simulated by Møller (1980) for half a year during 1975 by applying continuity considerations based on the observed in- and outflows through the Belts (Jacobsen 1980), which was estimated to account for 60% of the total flow. From these calculations an intermittency function can be outlined, where the intermittancy i is defined as the proportion of the total time in which the front is located north to the actual section. Assuming a Poisson distribution for i, we have

$$i = 1 - \exp\left(\frac{s-1,920 \cdot 10^3}{L_K}\right) \tag{16}$$

where $L_K = 40 \times 10^3$ m is the average inwards movement of the front.

In summary: In order to check the outlined salinity distribution with the salinities encountered in the Cattegat, we have to take the front movements into account, which yield an apparent salinity distribution

$$\overline{S_{KO}} = S_{K0} i + S_{K1}(1-i) \tag{17}$$

where $\overline{S_{KO}}$ is the yearly average salinity in the upper layer of the Cattegat. In Fig. 3 is shown the observed and the calculated yearly averaged salinities. The figure confirms the usefulness of the simple description, especially when it is realized that no »curve-fitting« is used in order to obtain agreement.

Subarea 2: The Belts

The hydrodynamics of subarea 2, which comprises the Belts, the Sound and the Kieler Bay are extremely complicated mainly due to the shallowness of the area (depth of 20 m to 30 m) combined with the highly non-steady in- and outflow of huge amounts of stratified brackish water. On an average a two-layer system exists. The existence of a salinity variation in the upper as well as in the lower layer shows that two-way entrainment is present, which agree with the lower layer being dynamical active in this region. A comprehensive description of the hydro-graphy (inclusive some considerations on the dynamic conditions) of subarea 2 can be found in The Belts project (1976), DHI report (1977) Bo Pedersen (1978) and Jacobsen (1980) where further references to the subject are given. Based on these findings (see also Fig. 2) we assume, that a typical flow cyclus representative of the present consequence analysis is as illustrated in Fig. 4, namely a net out-wards flow due to the fresh water discharge R super-imposed by an oscillatory motion with an amplitude in accordance with the measurements, i.e. approxima-tely a factor of ten times R. The amplitude and the frequency of the cyclus is governed by the meteorologic conditions as the forcing and the friction as the damping factors. As demonstrated by Bo Pedersen (1978), the total flow resistan-ce for the upper layer in the Great Belt is at its minimum at the present average

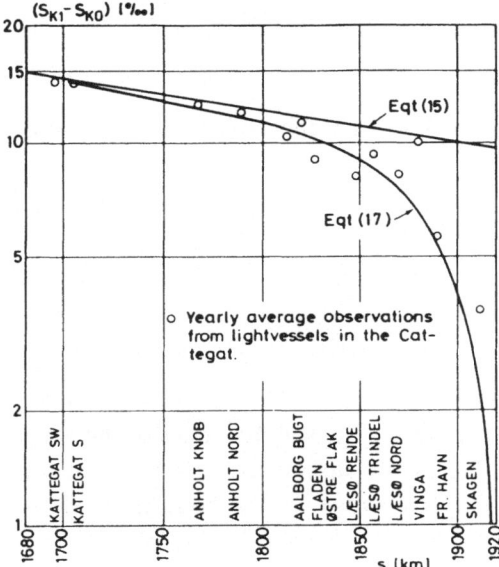

Fig. 3. Calculated and observed yearly average salinity differences between lower(S_{K1}) and upper (S_{K0}) layer in the Cattegat. Eq. (15) illustrates the calculated variation not taking the intermittancy into account (the front movements). Eq. (17) takes the intermittancy into account.

cyclic flow conditions, and hence a minor man-made change in the fresh water discharge will neither create changes in the amplitude nor in the frequency of the pendling discharge, see Fig. 4. Furthermore the minimum condition implies that no change in the production of turbulent kinetic energy occurs.

The sills which separate the Belts and the Sound from the Arkona Basin trap the inwards flowing water which descends as a dense bottom current into the lower layer of the stratified Bornholm Basin, see Fig. 1. If we assume, that the time in which trapping occur is nearly independent of the fresh-water discharge,

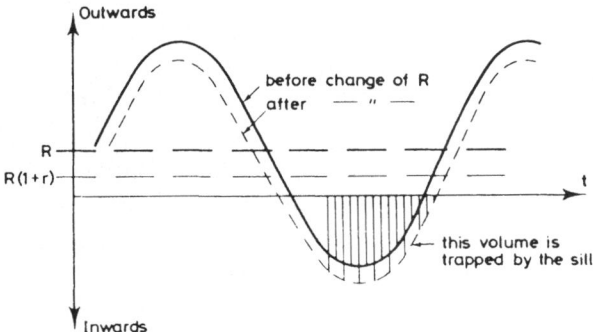

Fig. 4. The inflow/outflow through the Belts schematized by a simple harmonic cyclus superimposed on the fresh-water runoff. The volume trapped by the sills is hatched.

the following simple equation for the sill overflow Q_{SA1} (see Fig. 4) applies

$$Q_{SA1} + R = \text{const}_{SA1} \tag{18}$$

The other equations describing the model outlined are the continuity equations, which are

$$S_{K0}(Q_{ES0} - Q_{ES1} + Q_{SA1} + R) = S_{K1}(Q_{ES0} - Q_{ES1} + Q_{SA1}) \tag{19}$$

stating the no net transport of salt condition at the Belts/Cattegat transition, and

$$S_{SA1}Q_{SA1} = (Q_{ES0} - Q_{ES1} + Q_{SA1})S_{K1} - Q_{ES0}S_{S1} + Q_{ES1}S_{S0} \tag{20}$$

expressing the salt balance for the lower layer in the Belts. The entrainment fluxes (the Q_E's) are as stated in the previous chapter related to the dynamics of the flow, Eqs. (5) and (6). The salinities S_{S1} and S_{S0} in the upper and the lower layer, respectively are for conveniency taken as simple averages of the boundary values, i.e.

$$S_{S0} = 0.5(S_{KS0} + S_{SA0})$$
$$S_{S1} = 0.5(S_{KS1} + S_{SA1}) \tag{21}$$

The complexity of the flow pattern and of the geometry makes it a rather difficult task to give a convincing verification of the outlined Belt model, but a check can be made by considering some reasonable values for the measured/estimated quantities. Inserting the values shown in Fig. 1 in the governing equations yields

$$Q_{ES0} = 2.72 \cdot 10^4 \ m^3/s$$
$$Q_{ES1} = 2.18 \cdot 10^4 \ m^3/s \tag{22}$$

The upwards directed entrainment Q_{ES0} can be compared with the entrainment one would have had, if it was purely wind generated (see Eq. (3b)). With an interfacial area of approximately 10^{10} m^2 Eq. (3b) gives an upwards entraining flux of 1.2×10^4 m^3/s. The remaining entrainment, 1.5×10^4 m^3/s, can be accounted for by an upper layer dynamic velocity of $\bar{V} \simeq 0.45$ m/s (see Bo Pedersen 1980a) which is in good agreement with the prevailing flow conditions in the main contributor, the Great Belt. The downward entrainment is created by the turbulence production in the lower layer. The higher resistance the flow experienced at the bottom compared with the interface means, that the dynamic velocity of the lower layer necessary for producing the entrainment is less than that of the upper layer (approximately 2/3 of the upper layer velocity, see Bo Pedersen 1980a), which again agrees well with the observations in the Great Belt. Hence although the model is crude, it reflects the main hydrodynamic behaviour of the complex Belt-area.

Subarea 3: The Arkona Basin

The most characteristic hydrodynamic feature of the Arkona Basin can be observed during an inflow situation, where a dense bottom current is formed along the southern bank of the Basin. Passing the sills in the Great Belt and in the Sound the saline water flows into the Arkona Basin, where it very soon plunge down and descend to the deep water of the Bornholm Basin, i.e. below the interface located at approximately 45 m depth, see Fig. 1. During its course it entrains water from the less saline overlaying water, resulting in a decrease in salinity from 16‰ at the Darss Sill to about 11-12‰, at the merging with the lower layer in the Bornholm Basin. During a subsequent outflow situation the dense bottom current gradually runs dry and finally disappears until the process is repeated by a new inflow. The ambient water in the Arkona Basin is nearly homogeneous (except for the thermocline formation) because the average depth is less than the upper well-mixed layer of the adjacent Bornholm Basin. Hence the flow conditions in the upper layer of the Arkona Basin has no significant influence on the mixing of salt in the Baltic, and the salinity can therefore be taken as equal to the value present at the transition to the Bornholm Basin, namely 7.8‰.

The sparse measurements of the dense bottom current (Petrén et al. 1975), indicates an order of magnitude of typically 5 meter of the thickness of this current when present. If we apply Eq. (7) outlined in the previous chapter on the dense bottom current and takes the drop in elevation equal to the vertical distance from the plunge line to the interface in the Bornholm Basin ($y_{B0} = 45$ m) we get an average depth of $y_{A1} = 3.8$ m, in reasonable agreement with the observations. In these calculations we have applied the constant mass deficit flux condition

$$\Delta_{SA} Q_{SA1} = \Delta_B Q_{B1} \tag{23}$$

where indexes $SA1$ and $B1$ stand for the lower layers at the Darss Sill and at the Bornholm Basin respectively. A further condition necessary for the later calculations is Eq. (8) expressing the assumed constancy of the densimetric Froude number at the sill.

Subarea 4: The Bornholm Basin

The Bornholm Basin is a 80 to 100 m deep basin separated from the central Baltic Sea by a shallow sill with a narrow local depression to 60 m depth at the Stolpe Channel. The intermittent flow of dense water from the Arkona Basin is smoothed out and continues through the Stolpe Channel. Hence the lower layer in the Bornholm Basin only acts as a buffer for the discontinous inflow, with a retention time of say 10 weeks (Bo Pedersen 1977), while there is very little downwards directed mixing.

The position of the interface is determined by the rate of outflow through the Stolpe Channel. This flow is assumed to be governed by a constant densimetric Froude number, see Eq. (8). Hence the interfacial depth in the Bornholm Basin is related to the depth y_{C1} of the dense bottom current present in the Stolpe Channel and to the sill depth ($= 60$ m) in the following way

$$y_{B0} = 60\,(m) - y_{C1} \tag{24}$$

The upper layer is exposed to the wind and the heating/cooling sequence, which as stated in the previous chapter gives rise to an upwards directed entrainment through the halocline during half of the year. Hence Eq. (3b) shall be modified to

$$\frac{V_E}{U_F} = 0.5 \frac{2.3\,U_F^2}{\Delta g\,y_{B0}} \tag{25}$$

which yields a total discharge through the halocline of

$$Q_{EB0} = 1.5 \frac{U_F^3\,A_B}{g\,(S_{B1} - S_{B0})\,y_{B0}} \tag{26}$$

Inserting $U_F \simeq 8 \times 10^{-3}$ m/s yields $Q_{EB} = 8.9 \times 10^3$ m³/s in yearly average. The upper layer salinity S_{B0} can be found by applying the combined equation for transport of salt and volume through the Bornholm/Arkona section

$$S_{B0}\,(Q_{B1} + R) = S_{B1}\,Q_{B1} \tag{27}$$

A check on the calculated entrainment Q_{EB0} is that it is compatible with a salinity of $S_{\emptyset 0} = 7.0\%o$ in the upper layer of the Central Baltic, in accordance with observations.

Subarea 5: The Stolpe Channel
Surbarea 5 has the same hydrodynamic status as subarea 3, the Arkona Basin, the only significant difference being, that the dense bottom current here is nearly steady. The continuity of volume yields an upstream discharge of

$$Q_{BC1} = Q_{B1} - Q_{EB0} \tag{28}$$

i.e. the bottom water inflow to the Bornholm Basin Q_{B1} reduced by the upwards directed entrainment Q_{EB0}.

The salinity has accordingly decreased slightly, confer with continuity equation for mass or salt deficit

$$Q_{BC1}\,(S_{B1} - S_{\emptyset 0}) = Q_{\emptyset 1}\,(S_{\emptyset 1} - S_{\emptyset 0}) \tag{29}$$

which yields the salinity of inflowing bottom water to the Baltic Proper $S_{\emptyset 1} = 11\%o$ in agreement with observations.

Similar to the Arkona bottom current the depth of the Stolpe Channel current is assumed to be governed by a constant densimetric Froude number at the head of the current.

Subarea 6: The Central Baltic

The Central Baltic Basin is the largest of the subareas and furthermore the deepest (up to about 400 m). The main inflows of fresh water to the system takes place here, comprising direct river runoff (31%), contribution from the Finnish Bay (27%) and from the Bothnian Bay (42%).

The nearly persistent brackish water flow from the Stolpe Channel descends to below the primary halocline, located at approximately 60 m depth, where it spreads out at the density-matching level. The continuity of the volume demands an upwards directed entrainment $Q_{E\emptyset}$, caused by the energy input from the wind and the cooling, both of which are kept unchanged in the present analysis. The combined effects of all the external forces can be integrated to a single representative dynamic friction velocity $U_{F\emptyset}$, yielding an entrainment of (see Eq. (26))

$$Q_{E\emptyset} = Q_{\emptyset 1} = \frac{1.5 A_\emptyset U_{F\emptyset}^3}{(S_{\emptyset 1} - S_{\emptyset 0}) g y_{\emptyset 0}} \qquad (30)$$

Inserting the observed values gives a dynamic friction velocity of $U_{F\emptyset} \simeq 8 \times 10^{-3}$ m/s. Although this velocity compares well with the observed wind velocity, one has to remember that $U_{F\emptyset}$ contains the integrated effects of all the external forces and hence may be difficult to evaluate exactly, but as demonstrated, the order of magnitude is correct.

On an average, the upper layer of the Central Baltic is nearly homogeneous due to the multi-directed wind-driven circulations and the long retention time. The salinity is determined by the continuity equation for salt which states

$$S_{\emptyset 0}(R + Q_{\emptyset 1}) = S_{\emptyset 1} Q_{\emptyset 1} \qquad (31)$$

Subareas 7 and 8: The Finnish Bay and The Bothnian Bay

The salinity distribution and the position of the interface in the two subareas are governed by the boundary conditions, which, besides the external forces, are the fresh water input and the conditions in the Central Baltic, respectively. A change in these boundary conditions will have a measurable effect on the hydrography of both Bays' – but it will not influence the conditions in the Central Baltic – besides the changes caused by the change in the fresh-water input. Hence, as we are mainly concerned with the Central Baltic and the Danish Inland waters, a detailed analysis or modelling of the Finnish Bay and the Bothnian Bay is omitted. Another reason for not taking the two Bays into account is, that major changes may take place in the Finnish Bay which make the linearized approach doubtful.

The Consequence Analysis

The equations outlined in the previous chapters give an overall quantitative description of the Baltic estuary system, as it behaves under the present average meteorologic and hydrographic conditions. A man-made change in the runoff from the river Neva has of course a direct influence on the total fresh-water inflow to the Baltic, while it is unlikely to have any significant influence on the meteorological conditions and hence on the external forces. Therefore, the conditions before and after the river diversion both satisfies the outlined governing equations. Furthermore, the changes are relatively small, which suggests a linearization of the equations. The procedure is to introduce the new parameters as the old ones plus a minor correction, as for instance

$$
\begin{aligned}
R_{new} &= R(1 + r) \\
Q_{new} &= Q(1 + q) \\
S_{new} &= S(1 + s) \\
y_{new} &= y(1 + \eta) \\
\lambda_{new} &= \lambda(1 + \ell)
\end{aligned}
\tag{32}
$$

where r, q, s, η and λ all are small dimensionless quantities. By use of a Taylor expansion in which only the first order terms are retained, a set of linear equations in the correction terms is obtained, which can be solved directly. The original as well as the linearized equations are for conveniency summarized in Table 1, where the solution to the set of equations is shown as well. In the linearized equations due respect to the changes in the interfacial widths or areas with depth have been taken.

All the corrections have been related to the fresh-water discharge reduction r in the data output. Hence – as an example – if the fresh-water diversion from the river Neva amounts to 5 per cent of the total fresh-water inflow to the Baltic (i.e. approximately 25 per cent reduction of the river Neva's discharge), then $r = -0.05$. By use of Table 1 the associated changes in the salinities, depths and discharges can be evaluated. Some of the calculated changes can be compared with the natural variations encountered in the Baltic estuary system during this century. We have chosen to illustrate this variability by plotting the 10-year sliding mean values of the runoff R from river Vuoksi, the salinity $S_{\emptyset 0}$ and the depth $y_{\emptyset 0}$ in the upper layer in the Central Baltic and finally the upper layer salinity in the Cattegat region, see Fig. 5. Two comments to this illustration are appropriate. First, the natural variations are caused by the combined effects of a variability in the runoff and in the climate, the last one being held unchanged in our calculations. Second, the natural variations are highly non-steady, while our analysis deals with the steady state. In the non-steady case, the reservoir effect damps the amplitudes of a cyclic variation (as for instance in the salinity) compared to the long-term response of a step function and furthermore the output is delayed compared to the time

TABLE 1

EQT. No	THE GOVERNING EQUATIONS	REGION	THE LINEARIZED EQUATIONS	SOLUTION
14	$B^2 \nu_{K0}^3 (33 \cdot 10^{-3} - S_{MK0}) = 9.44 \cdot 10^{10}$	K	$1.80 n_{K0} - 2.37 s_{MK0} = 0$	$n_{K0} = -0.36\,r$
18	$(33 \cdot 10^{-3} - S_{MK0}) = (33 \cdot 10^{-3} - S_{KS0}) \exp(-\frac{240 \cdot 10^3 m}{\lambda_K})$		$-2.37 s_{MK0} + 1.20 s_{KS0} - 0.43 \ell_K = 0$	$s_{MK0} = -0.27\,r$
16	$\lambda_K = 1.81 \cdot 10^5 \frac{R \nu_{K0}}{B_K}$		$-\ell_K + 1.60 n_K = -r$	$s_{KS0} = -0.39\,r$
	$B_K = 64 \cdot 10^3 (1 - 0.6 n_{K0})$			$\ell_K = 0.42\,r$
21	$Q_{SA1} + R = 2.92 \cdot 10^4 \,[m^3/s]$	S	$-1.42 q_{SA1} = 1.50\,r$	$q_{SA1} = -1.06\,r$
22	$S_{KS0}(Q_{ES0} - Q_{ES1} + Q_{SA1} + R) =$ $33 \cdot 10^{-3}(Q_{ES0} - Q_{ES1} + Q_{SA1})$		$s_{KS0} - 0.69 q_{ES0} + 0.59 q_{ES1} - 0.36 q_{SA1} = -0.46\,r$	$q_{ES0} = -0.03\,r$ $q_{ES1} = -0.81\,r$
23	$S_{SA1} Q_{SA1} = (Q_{ES0} - Q_{ES1} + Q_{SA1}) 33 \cdot 10^{-3}$ $- Q_{ES0} S_{S1} + Q_{ES1} S_{S0}$		$s_{SA1} - 1.06 q_{SA1} - 1.02 q_{ES0} + 2.07 q_{ES1}$ $+ 2.93 s_{S1} - 1.33 s_{S0} = 0$	$s_{SA1} = 0.04\,r$ $s_{S1} = 0.01\,r$
24	$S_{S0} = 0.5(S_{KS0} + S_{B0})$ $S_{S1} = 0.5(33 \cdot 10^{-3} + S_{SA1})$		$s_{S0} - 0.70 s_{KS0} - 0.30 s_{B0} = 0$ $s_{S1} - 0.33 s_{SA1} = 0$	$s_{S0} = -0.33\,r$
8	$Q_{ES0} = \frac{4.20 \cdot 10^3}{(S_{S1} - S_{S0}) \nu_{K0}}$		$q_{ES0} + 1.0 n_{K0} + 2.11 s_{S1} - 1.11 s_{S0} = 0$	
9	$Q_{ES1} = \frac{3.22 \cdot 10^3}{(S_{S1} - S_{S0})(25 - \nu_{K0})}$		$q_{ES1} - 1.14 n_{K0} + 2.11 s_{S1} - 1.11 s_{S0} = 0$	
26	$(S_{SA1} - S_{B0}) Q_{SA1} = (S_{B1} - S_{B0}) Q_{B1}$	A	$1.95 s_{SA1} + 1.28 s_{B0} + q_{SA1} - q_{B1} - 3.23 s_{B1} = 0$	
10	$Q_{B1} = Q_{SA1} \exp\{0.072 \frac{\nu_{B0}}{\nu_{A1}}\}$		$q_{B1} - q_{SA1} + 0.85 n_{A1} - 0.85 n_{B0} = 0$	$n_{A1} = -0.79\,r$
11	$\frac{Q_{SA1}^2}{(S_{SA1} - S_{B0}) \nu_{A1}^3} = 4.48 \cdot 10^8$		$2 q_{SA1} - 1.95 s_{SA1} + 0.95 s_{B0} - 3 n_{A1} = 0$	
27	$\nu_{B0} = 60 [m] - \nu_{C1}$	B	$3 n_{B0} + n_{C1} = 0$	$s_{B0} = 0.16\,r$ $s_{B1} = -0.31\,r$
29	$Q_{EB0} = \frac{7.79 \cdot 10^{-8} A_B}{(S_{B1} - S_{B0}) \nu_{B0}}$		$q_{EB0} + 1.60 n_{B0} + 3.23 s_{B1} - 2.23 s_{B0} = 0$	$n_{B0} = 0.16\,r$
30	$S_{B0}(Q_{B1} + R) = S_{B1} Q_{B1}$ $A_B = 1.8 \cdot 10^{10}(1 - 0.6 n_{B0})$		$s_{B0} - s_{B1} + 1.69 q_{B1} = -0.31\,r$	$q_{B1} = -0.25\,r$ $q_{EB0} = 0.29\,r$
31	$Q_{BC1} = Q_{B1} - Q_{EB0}$	C	$q_{BC1} - 1.36 q_{B1} + 0.36 q_{EB0} = 0$	$n_{C1} = -0.48\,r$
32	$Q_{BC1}(S_{B1} - S_{\emptyset 0}) = Q_{\emptyset 1}(S_{\emptyset 1} - S_{\emptyset 0})$		$q_{BC1} - q_{\emptyset 1} + 2.63 s_{B1} + 0.12 s_{\emptyset 0} - 2.75 s_{\emptyset 1} = 0$	$q_{BC1} = -0.44\,r$
11	$\frac{Q_{BC1}^2}{(S_{B1} - S_{\emptyset 0}) \nu_{C1}^3} = 4.10 \cdot 10^7$		$2 q_{BC1} - 2.63 s_{B1} + 1.63 s_{\emptyset 1} - 3 n_{C1} = 0$	
10	$Q_{\emptyset 1} = Q_{BC1} \exp\{0.072 \frac{\nu_{\emptyset 0} - \nu_{B0}}{\nu_{C1}}\}$		$q_{\emptyset 1} - q_{BC1} - 0.29 n_{\emptyset 0} + 0.22 n_{B0} + 0.07 n_{C1} = 0$	
33	$Q_{\emptyset 1} = \frac{7.89 \cdot 10^{-8} A_\emptyset}{(S_{\emptyset 1} - S_{\emptyset 0}) \nu_{\emptyset 0}}$	Ø	$q_{\emptyset 1} + 2.1 n_{\emptyset 0} + 2.75 s_{\emptyset 1} - 1.75 s_{\emptyset 0} = 0$	$q_{\emptyset 1} = -0.46\,r$ $s_{\emptyset 0} = -0.85\,r$
34	$S_{\emptyset 0}(R + Q_{\emptyset 1}) = S_{\emptyset 1} Q_{\emptyset 1}$ $R = 1.5 \cdot 10^4 m^3/s$ $A_\emptyset = 8 \cdot 10^{10}(1 - 1.1 n_{\emptyset 0})$		$s_{\emptyset 0} - 0.36 q_{\emptyset 1} - s_{\emptyset 1} = -0.36\,r$	$s_{\emptyset 1} = -0.33\,r$ $n_{\emptyset 0} = -0.06\,r$

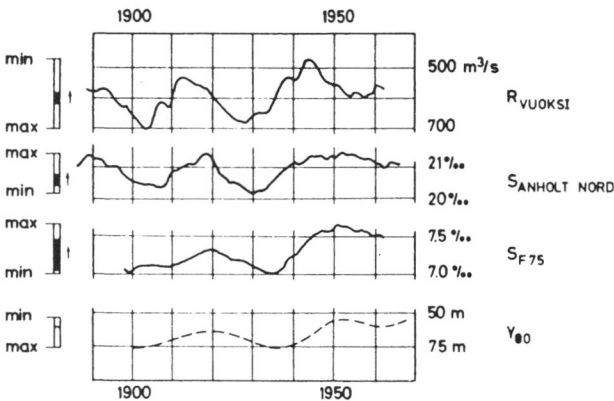

Fig. 5. The secular changes in the Baltic Sea estuary system illustrated by the 10-year sliding mean of the runoff from river Vuoksi (Nilsson and Svansson 1974), the surface salinity at Anholt Nord (Nilsson and Svansson 1974), the upper layer salinity at station F75 (the Central Baltic, Hela 1966), the upper layer depth in the Central Baltic (Fonselius 1969).

In the column diagram to the left is shown the min/max values observed and the changes calculated for a 25 per cent reduction of the river Neva's runoff.

for the input. The time delaying effect is demonstrated in the salinity observed at station F75, which shows a response time of about 10 years, compatible with the retention time in the upper layer of the Central Baltic. A throughout discussion of the observed data shall not be given, mostly due to lack of knowledge of the variability in the meteorologic forcing. Instead we shall compare and comment upon the natural and man-made variations as observed and calculated, respectively.

The man-made reduction in our example is 5% of the total runoff to the Baltic. This reduction is an order of magnitude less than the natural variations encountered during this century as represented by the runoff of river Vuoksi, see Fig. 5.

The most pronounced influence of a man-made reduction in the runoff is to be found in the upper layer salinity in the Central Baltic, see Fig. 5, S_{F75}, where the calculated salinity change is approximately half the variation observed during this century. The observed salinity variations reflects the non-steady input from fresh water (R) and wind (U_F), and are therefore highly damped by the reservoir effect, as compared to the values one would have had if a permanent change in R and U_F had occurred. Moreover the observed salinity is highly influenced by the change in the meteorological conditions, which are not taken into account in our calculations. Therefore, the relatively strong influence of a man-made change in the runoff is understandable.

The change in salinity of the upper layer of the Cattegat light-vessel Anholt Nord is only approximately 20% of the natural maximum variation observed. In fact the 10-year sliding mean is a bad representation of the salinities in the Cattegat region because the retention time is orders of magnitude less. Compared to the more representative month sliding mean the influence is reduced to less than say 5% of the natural variations.

Finally we found a negligible change in the upper layer depth of the Central Baltic. The actual maximum change has been about 20%. Hence the position of the interface in the Central Baltic is primarily determined by the changes in the meteorological conditions, which have an influence directly in determining the rate of entrainment and indirectly in determining the sill overflow and hence the discharge into the lower layer.

Conclusion

The possible influence on the hydrography of the Baltic Sea estuary system subject to a man-made change in the river runoff, has been investigated. The density stratification, the geometry and the variability of the external forces makes the estuary a rather complex hydrodynamic system and hence it was necessary to divide it into eight subareas, each of which described by its own balance in mass, volume and dynamic (mixing). It has been demonstrated that the prevailing meteorological conditions over the area play the dominating role in the formation of fronts, salt water wedges, dense bottom currents and all the other types of density currents encountered in the system and in the mixing processes. These meteorological conditions are kept unchanged in the present consequence analysis, which has been performed by linearizing the governing equations with respect to the minor changes caused by the man-made change in the river Neva's runoff. The changes in the depths, discharges and salinities in the system have been presented as functions of the change in the fresh-water discharge. Although it is not the intention of this paper to reach any conclusions concerning the possible positive or negative consequences of the river diversion it shall be emphasized, that the dynamic stability of the estuary system will be reduced, which makes the system more sensitive to the inevitable changes in the external forces, i.e. in the natural variations in the meteorology. This reduced stability is caused by the combined effects of an increased salinity and a decreased upper layer depth.

The salinity increase in the upper layer of the Central Baltic is remarkably large compared to the natural variations encountered during this century. The reason for this is the reservoir effect, which highly damps a cyclic variation (the natural) but not a step-variation (the man-made).

References

Bo Pedersen, Fl. (1977) On dense bottom currents in the Baltic Deep Water. *Nordic Hydrology,* 8, 297-316.

Bo Pedersen, Fl. (1978) On the influence of a bridge across the Great Belt on the hydrography of the Baltic Sea, 11th conference of the Baltic Oceanographers, Rostock, DDR.

Bo Pedersen, Fl. (1980a) A monograph on turbulent entrainment and friction in two-layer stratified flow. Series Paper No. 25, Inst. of Hydrodynamics and Hydraulic Eng., Tech. Univ. of Denmark.

Bo Pedersen, Fl. (1980b) Dense bottom currents in Rotating Ocean. American Society of Civil Engineers, *Proc. Vol. 106, Hy 8,* 1291-1308.

DHI-Report (1977) Bæltprojektet. Matematiske modeller af Store Bælt og Øresund – Slutrapport. Dansk Hydraulisk Inst. DK-2970 Hørsholm.

Fonselius, S. H. (1969) Hydrography of the Baltic Deep Basins III, Fishery Board of Sweden, Series Hydrography, Report No. 23.

Golubev, G. (1978) Environmental Issue of Large Interregional Water Transfer Projects. *Water Supply and Management, Vol. 2,* 177-185.

Hela, I. (1966) Secular changes in the salinity of the upper waters of the Northern Baltic Sea. *Commentationes Physico-Mathematicae, Vol. 31,* Nr. 14, 1966.

Jacobsen, T. (1980) Sea Water exchange of the Baltic. Measurements and methods. Preprint of Dr. thesis. Inst. of Physical Oceanography. Univ. of Copenhagen.

Mikhaylov, N. I., Nikolayev, V. A., and Timashev, I. Ye. (1977) Environmental Protection Issue and Southward Diversion of Siberian Rivers. English Translation from: Vestnik Moskovskog Universiteta, Geografiya (1977), No. 5, pp. 50-56.

Møller, Jacob Steen (1980) Østersøens Hydrografi. Internal Report in Danish. Inst. of Hydrodynamics and Hydraulic Eng. Tech. Univ. of Denmark.

Nilsson, H., and Svansson, A. (1974) Long term variations of Oceanographic Parameters in the Baltic and Adjacent Waters. Meddelande från Havsfiske-laboratoriet, Lysekil, nr. 174.

Petrén, O., and Walin, G. (1975) Some observations of the deep flow in the Bornholm strait during the period June 73 – December 74, Rep. No. 12, Inst. of Oceanography, Univ. of Gothenburg, Sweden.

Rydberg, L. (1976) Observations of the deep water flow through the Stolpe Channel during August 1976. Rep. No. 15. Inst. of Oceanography, Univ. of Gothenburg, Sweden.

The Beltproject (1976) Interim report on the Danish Belt project (In Danish). Publ. by Miljøstyrelsen, Kampmannsgade 1, DK-1604 Copenhagen.

Voropaev, G. V. (1978) The Scientific Principles of Large-Scale Areal Redistribution of Water Resources in the USSR. *Water Supply and Management, Vol. 2,* pp. 91-101.

Received: 18 December, 1980

Address:
Institute of hydrodynamics and hydraulic engineering,
ISVA,
Technical University of Denmark,
Building 115,
DK-2800 Lyngby, Denmark.

March, 1973

Journal of the
HYDRAULICS DIVISION
Proceedings of the American Society of Civil Engineers

SURFACE JET AT SMALL RICHARDSON NUMBERS

By Frank Engelund[1] and Flemming Bo Pedersen[2]

INTRODUCTION

Spreading of a jet discharged horizontally at the surface of an initially quiescent water of larger density has been investigated by several authors (2,3,4). The density difference may be due to a higher temperature of the jet, or a larger salinity of the ambient fluid, or both. In the present paper it has been attempted to develop a theoretical solution based on some assumptions of similarity in velocity and density distributions in the case of small Richardson numbers.

One important feature of this flow is the entrainment of ambient fluid. The present knowledge of this process seems to be rather incomplete, but some rational information has been achieved by Ellison and Turner (1).

BASIC ASSUMPTIONS

The equations developed herein make use of a horizontal coordinate system with the x_1-axis in the jet axis (see Fig. 1). The theory is based on the following main assumptions:

1. After a short zone of flow establishment near the outlet, the flow attains a state characterized by general similarity of the velocity and density profiles. A typical example of a velocity profile is given in Fig. 2(c). The exact form of the profile is not known and is, in fact, of minor importance for the theory.

2. According to measurements by Hayashi and Shuto (2) and Wilkinson (6), the variation of mean density along a vertical is nearly linear, as indicated in

Note.—Discussion open until August 1, 1973. To extend the closing date one month, a written request must be filed with the Editor of Technical Publications, ASCE. This paper is part of the copyrighted Journal of the Hydraulics Division, Proceedings of the American Society of Civil Engineers, Vol. 99, No. HY3, March, 1973. Manuscript was submitted for review for possible publication on May 10, 1972.

[1] Prof. of Hydr., Inst. of Hydrodynamics and Hydr. Engrg., Tech. Univ. of Denmark, Copenhagen, Denmark.
[2] Research Engr., Inst. of Hydrodynamics and Hydr. Engrg., Tech. Univ. of Denmark, Copenhagen, Denmark.

Fig. 2(a). In the following such a linear variation is adopted as one of the basic assumptions and the interface is defined as the level where the density of the jet fluid equals the density, ρ_o, of the ambient fluid.

3. The densimetric Froude number is defined as

$$F_\Delta = \left(\frac{v^2}{\Delta g S}\right)^{1/2} \quad \dots\dots\dots\dots\dots\dots\dots\dots\dots\dots\dots\dots \quad (1)$$

in which v = the surface velocity; g = the acceleration of gravity; and S = the

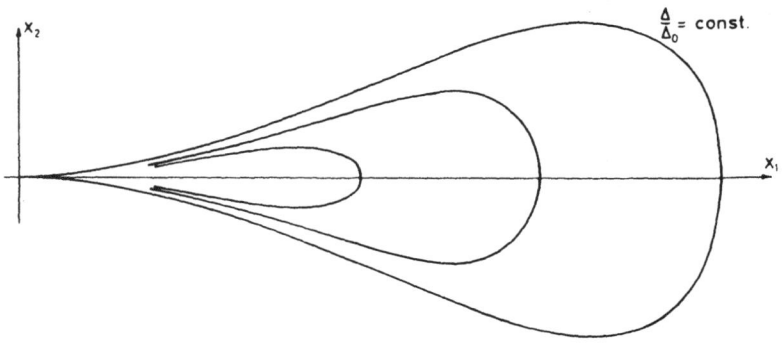

FIG. 1.—PLAN VIEW AND DENSITY CONTOURS

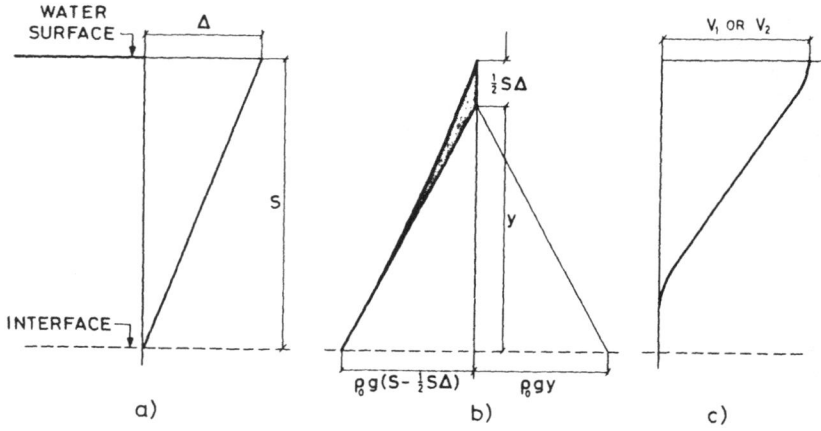

FIG. 2.—(a) DISTRIBUTION OF DENSITY DEFECT; (b) PRESSURE DISTRIBUTION IN JET (LEFT) AND AMBIENT FLUID (RIGHT); (c) VELOCITY DISTRIBUTION

vertical thickness of the jet, measured from the surface to the interface (see Fig. 3). The quantity, Δ, is defined by

$$\Delta = \frac{\rho_o - \rho}{\rho_o} \quad \dots\dots\dots\dots\dots\dots\dots\dots\dots\dots\dots\dots\dots\dots \quad (2)$$

in which ρ = the density of the jet fluid at the water surface.

The densimetric Froude number, F_Δ, is assumed to be large near the outlet and to decrease in the flow direction. Value F_Δ approaches unity far downstream, but the theory fails to account for the further development of the flow after F_Δ about 3.

4. Shear stresses in vertical planes are assumed to be negligible. In horizontal planes, on the other hand, the shear stresses are not negligible, but they will not appear explicitly in the integrated equations developed herein. According to Pedersen (5), the distribution of the horizontal shear stress is parabolic, and no shear is transferred from the jet to the ambient fluid.

5. The pressure distribution along any vertical is assumed to be hydrostatic. However, because the density varies vertically, this does not imply a linear, but a parabolic variation of the pressure, as indicated in Fig. 2(*b*).

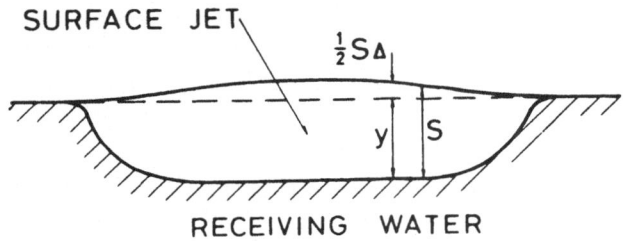

FIG. 3.—CROSS SECTION OF SURFACE JET

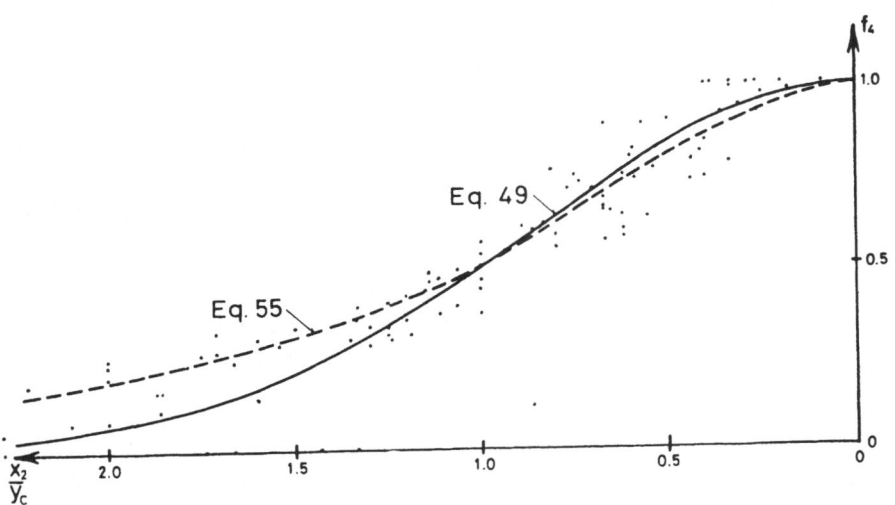

FIG. 4.—COMPARISON BETWEEN THEORETICAL AND EXPERIMENTAL (3) TRANSVERSE DISTRIBUTION OF DENSITY DEFICIT

If S denotes the local thickness of the jet and y is the corresponding difference in level between the interface and the surface of the ambient water, the following hydrostatic relation is obtained:

$$\left(\rho_o - \frac{1}{2}\,\Delta\,\rho_o\right) S = \rho_o y \quad \dots\dots\dots\dots\dots\dots\dots\dots\dots \quad (3)$$

or simply $S - y = \frac{1}{2}\,S\Delta$ $\dots\dots\dots\dots\dots\dots\dots\dots\dots\dots\dots$ (4)

which is the superelevation of the jet surface over the ambient water level. Thus, the total difference in pressure per unit width over the vertical amounts to

$$\frac{1}{6} \ \Delta g S^2 \quad \dots\dots\dots\dots\dots\dots\dots\dots\dots\dots\dots\dots\dots\dots\dots \quad (5)$$

corresponding to the shaded area in Fig. 2(b).

6. The relative density difference is so small that its influence on the inertia in negligible (Boussinesq approximation).

7. In the case of miscible fluids considered herein, a crucial point is the entrainment of ambient fluid into the jet. The entrainment is a one-way process not to be confused with turbulent diffusion. The entrainment velocity, E, is defined as the rate of entrainment per unit area and may, according to several investigations, be given by a relation of the type

$$\frac{E}{v} = f(F_\Delta) \quad \dots\dots\dots\dots\dots\dots\dots\dots\dots\dots\dots\dots\dots\dots \quad (6)$$

in which the entrainment function, f, has to be found empirically (1). Thus, for a surface jet, f varies longitudinally. As the depth to width ratio is low the entrainment at the sides of the jet will be neglected.

Since large values of F_Δ indicate that inertia dominates over buoyancy, Ellison and Turner (1) assumed that f approaches a constant value for increasing densimetric Froude number.

This constant value must be the one pertaining to two-dimensional jets in homogeneous fluids, i.e., in the absence of density differences, as pointed out by Ellison and Turner (1), who suggest the value

$$f = 0.075 \quad \dots\dots\dots\dots\dots\dots\dots\dots\dots\dots\dots\dots\dots\dots\dots \quad (7)$$

when a velocity near the surface is applied in Eq. 6.

The information available on the entrainment factor is not very detailed for large F_Δ, but it seems that little deviation from the value given in Eq. 7 is likely to occur before F_Δ becomes so small that the assumption on which the present theory is based will no longer be fulfilled.

8. Heat exchange with the atmosphere has been neglected for the case of heated discharges.

DERIVATION OF FLOW EQUATIONS

According to the assumptions stated previously, only the two horizontal coordinates, x_1 and x_2, are needed. As the flow is assumed to be steady, the time does not appear in the equations.

Equation of Continuity.—The rate of flow through a rectangular section of unit width perpendicular to the x_1-axis is

$$\alpha v_1 S \quad \dots\dots\dots\dots\dots\dots\dots\dots\dots\dots\dots\dots\dots\dots\dots \quad (8)$$

in which v_1 = the surface velocity component; and α = a correction factor. The profile measured by Wilkinson (6) indicates that α is of the order of one-third.

In the following a vertical column of the jet is considered, the height being equal to the thickness, S, while the base is a unit area. The continuity equation reads

$$\frac{\partial}{\partial x_1} (\alpha v_1 S) + \frac{\partial}{\partial x_2} (\alpha v_2 S) = \frac{\partial}{\partial x_i} (\alpha v_i S) = E \dots \dots \dots \dots \quad (9)$$

the left-hand side being the net outflow of fluid through the cylinder surface and the right-hand side being the rate of entrainment through the base. The middle term is, of course, merely the left hand side written in Cartesian tensor notation with $i = 1$ and 2.

Equation of Mass Conservation.—In order to express the conservation of mass the previously defined cylinder is again used as a control surface and thus

$$\rho_o E = \frac{\partial}{\partial x_i} \int_0^S (\rho_o - \Delta \rho_o) \, v_i \, dx_3 \dots \dots \dots \dots \dots \quad (10)$$

in which the left-hand side is the rate of mass entrainment through the base, while the term on the right-hand side is the net rate of mass flux through the cylinder surface.

When Eqs. 2 and 9 are inserted in Eq. 10, the following equation for conservation of mass deficit is obtained:

$$\frac{\partial}{\partial x_i} (\Delta v_i S) = 0 \dots \dots \dots \dots \dots \dots \dots \dots \dots \quad (11)$$

Momentum Equation.—The longitudinal momentum flux through an area of unit width perpendicular to the x_1-axis is

$$\beta \rho v_1^2 S \approx \beta \rho_o v_1^2 S \dots \dots \dots \dots \dots \dots \dots \dots \dots \quad (12)$$

in which β is a nondimensional correction factor. Wilkinson's velocity profile (6) gives a value of the order of $\beta \sim 0.2$.

Applying the usual control surface, the momentum equation becomes

$$\frac{\partial}{\partial x_j} (\beta v_i v_j S) = - \frac{\partial}{\partial x_i} \left(\frac{1}{6} \Delta g S^2 \right) \dots \dots \dots \dots \dots \dots \quad (13)$$

the left-hand side being the net momentum flux and the right-hand side being the resulting pressure force, e.g., Eq. 5. In principle, Eqs. 9, 11, and 13 determine the four dependent variables v_1, v_2, S, and Δ.

Nondimensional Form of Equations.—For the subsequent analysis it is convenient to apply dimensionless variables. To this end, a characteristic velocity, v_o, and a relative density difference, Δ_o, are introduced, both, for instance, being maximum values at a certain reference section. Then the following nondimensional variables may be introduced:

$$u_i = \frac{v_i}{v_o} \text{ and } \frac{\Delta}{\Delta_o} \dots \dots \dots \dots \dots \dots \dots \dots \dots \quad (14)$$

Correspondingly, the symbol, x_i, will be used subsequently to denote the true coordinates divided by a horizontal length scale, L. For the depth, however, a vertical length scale, D, is used, so that

$$s = \frac{S}{D} \dots \dots \dots \dots \dots \dots \dots \dots \dots \dots \dots \dots \quad (15)$$

in which D may be taken as the maximum depth in the reference section.

Further, the dimensionless parameters are defined as

$$R = \frac{\Delta_0 g D}{6\beta v_o^2} \quad \text{(Richardson number)} \quad \dots \dots \dots \dots \dots \dots \dots \quad (16)$$

$$e = \frac{E}{\alpha v_1} \quad \text{(entrainment factor)} \quad \dots \dots \dots \dots \dots \dots \dots \dots \quad (17)$$

Then the flow equations become

$$\frac{\partial}{\partial x_j} (u_i u_j s) = - R \frac{\partial}{\partial x_i} \left(\frac{\Delta}{\Delta_o} s^2 \right) \quad \dots \dots \dots \dots \dots \dots \quad (18)$$

$$\frac{\partial}{\partial x_i} (u_i s) = e \frac{L}{D} u_1 \dots \dots \dots \dots \dots \dots \dots \dots \dots \dots \quad (19)$$

$$\frac{\partial}{\partial x_i} \left(\frac{\Delta}{\Delta_o} u_i s \right) = 0 \quad \dots \dots \dots \dots \dots \dots \dots \dots \dots \dots \quad (20)$$

SOLUTION OF EQUATIONS

The assumption of a small Richardson number implies that the densimetric Froude number, F_Δ, is large. Physically, this means that the longitudinal momentum is dominating as compared with pressure, so that the pressure term may be dropped. However, in the transverse direction pressure must be of an order of magnitude comparable to the momentum, because the spreading of the jet depends mainly on the increased pressure in the middle of the jet. In these circumstances, the flow equations reduce to

$$\frac{\partial}{\partial x_1} (u_1^2 s) + \frac{\partial}{\partial x_2} (u_1 u_2 s) = 0 \quad \dots \dots \dots \dots \dots \dots \dots \quad (21)$$

$$\frac{\partial}{\partial x_2} (u_2^2 s) + \frac{\partial}{\partial x_1} (u_1 u_2 s) = - R \frac{\partial}{\partial x_2} \left(\frac{\Delta}{\Delta_o} s^2 \right) \quad \dots \dots \dots \dots \quad (22)$$

$$\frac{\partial}{\partial x_1} (u_1 s) + \frac{\partial}{\partial x_2} (u_2 s) = e \frac{L}{D} u_1 \quad \dots \dots \dots \dots \dots \dots \dots \quad (23)$$

$$\frac{\partial}{\partial x_1} \left(\frac{\Delta}{\Delta_o} u_1 s \right) + \frac{\partial}{\partial x_2} \left(\frac{\Delta}{\Delta_o} u_2 s \right) = 0 \quad \dots \dots \dots \dots \dots \dots \quad (24)$$

One way of obtaining a solution to such a system is to make similarity assumptions which allow the transformation of partial differential equations into ordinary differential equations. In order to do so, the variable

$$\xi = \frac{x_2}{B} \quad \dots \dots \dots \dots \dots \dots \dots \dots \dots \dots \dots \dots \dots \dots \dots \dots \quad (25)$$

is introduced, in which B measures the lateral extension of the jet. Further, it is assumed that the dependent variables may be written in the following way:

$$u_1 = x_1^{-a} f_1(\xi) \quad \dots \dots \dots \dots \dots \dots \dots \dots \dots \dots \dots \dots \dots \quad (26)$$

$$u_2 = \sqrt{R} \, x_1^{-b} f_2(\xi) \quad \dots \dots \dots \dots \dots \dots \dots \dots \dots \dots \dots \quad (27)$$

$$s = x_1^c f_3(\xi) \quad \dots \dots \dots \dots \dots \dots \dots \dots \dots \dots \dots \dots \dots \quad (28)$$

$$\frac{\Delta}{\Delta_o} = x_1^{-d} f_4(\xi) \quad \dots \dots \dots \dots \dots \dots \dots \dots \dots \dots \dots \dots \quad (29)$$

$$B = \sqrt{R} \; x_1^n \dots \dots \dots \dots \dots \dots \dots \dots \dots \dots \dots \dots \dots \quad (30)$$

The exponents a, b, c, d, and n are ordinary numbers and the f terms are unknown functions. The characteristic velocity, v_o, being defined as the maximum in the reference section, it follows from Eq. 14 that $u_1 = 1$ for $\xi = 0$ in this section. If f_1 is normalized so that $f_1(0) = 1$, thus, from Eq. 26, the reference section must correspond to $x_1 = 1$. Similarly, it is seen that $f_4(0) = 1$ may be taken for $x_1 = 1$.

For large values of the densimetric Froude number the entrainment function is constant, as mentioned in item 7. For the determination of the five unknown exponents five equations are needed. The first three are obtained if Eqs. 26-30 are substituted into Eqs. 21-24 and if all terms are required to be varying with the same power:

$$- a + b + n = 1 \dots \dots \dots \dots \dots \dots \dots \dots \dots \dots \quad (31)$$

$$2b + c - d = 0 \dots \dots \dots \dots \dots \dots \dots \dots \dots \dots \quad (32)$$

$$c = 1 \dots \dots \dots \dots \dots \dots \dots \dots \dots \dots \dots \dots \dots \quad (33)$$

The last two equations come from the conditions that the total longitudinal momentum as well as the total mass deficit flux must be constant. This gives the following equations

$$- 2a + c + n = 0 \dots \dots \dots \dots \dots \dots \dots \dots \dots \dots \quad (34)$$

$$- a + c - d + n = 0 \dots \dots \dots \dots \dots \dots \dots \dots \dots \quad (35)$$

The solution of these linear equations yields $a = d = 5/3$; $b = 1/3$; $c = 1$; and $n = 7/3$.

Applying these values, the flow equations reduce to the following four simultaneous differential equations for the unknown functions:

$$\frac{d}{d\xi} \, (f_1 f_2 f_3) = \frac{7}{3} \, f_1^2 f_3 + \frac{7}{3} \, \xi \, \frac{d}{d\xi} \, (f_1^2 f_3) \dots \dots \dots \dots \dots \quad (36)$$

$$- \frac{d}{d\xi} \, (f_2^2 f_3) + f_1 f_2 f_3 + \frac{7}{3} \, \xi \, \frac{d}{d\xi} \, (f_1 f_2 f_3) = \frac{d}{d\xi} \, (f_4 f_3^2) \dots \dots \dots \quad (37)$$

$$\frac{2}{3} \, f_1 f_3 + \frac{7}{3} \, \xi \, \frac{d}{d\xi} \, (f_1 f_3) = \frac{d}{d\xi} \, (f_2 f_3) - e \, \frac{L}{D} \, f_1 \dots \dots \dots \dots \quad (38)$$

$$\frac{7}{3} \, f_1 f_3 f_4 + \frac{7}{3} \, \xi \, \frac{d}{d\xi} \, (f_1 f_3 f_4) = \frac{d}{d\xi} \, (f_2 f_3 f_4) \dots \dots \dots \dots \quad (39)$$

Even though this system of equations appears somewhat complicated, it is, in fact, rather easy to obtain a simple exact solution, as demonstrated in the following. First, it is realized that the only possible solution of Eq. 39 is

$$f_2 = \frac{7}{3} \, \xi \, f_1 \dots \dots \dots \dots \dots \dots \dots \dots \dots \dots \dots \dots \quad (40)$$

Inserted in Eq. 38 the latter reduces to

$$f_3 = \frac{3}{5} \, \frac{L}{D} \, e \dots \dots \dots \dots \dots \dots \dots \dots \dots \dots \dots \dots \quad (41)$$

which shows that the jet thickness is constant in a given cross section. Since the vertical length scale, D, is defined as the thickness for $x_1 = 1$, it is seen from Eq. 28 that $f_3 = 1$, so that

$$e = \frac{5}{3} \frac{D}{L} \dots\dots\dots\dots\dots\dots\dots\dots\dots\dots \quad (42)$$

Experiments by Jen and Wiegel (3) confirm that the jet thickness, S, varies linearly with the true distance, $x_1 L$, according to the expression

$$S = 0.15 \ (x_1 L) \dots\dots\dots\dots\dots\dots\dots\dots\dots \quad (43)$$

or $\ s = \frac{S}{D} = 0.15 \ x_1 \frac{L}{D} \dots\dots\dots\dots\dots\dots\dots\dots \quad (44)$

which agrees with $f_3 = 1$ if

$$D = 0.15 \ L \dots\dots\dots\dots\dots\dots\dots\dots\dots\dots \quad (45)$$

Referring to Eqs. 17 and 42, the entrainment factor becomes $E/v_1 = \alpha e = 0.25 \ \alpha$. This is in agreement with the entrainment factor mentioned previously if α is taken to be 0.3, which is the right order of magnitude.

Substituting Eq. 40 and $f_3 = 1$ in Eq. 36, this is seen to be identically satisfied. This means that the system of equations degenerates, so that an additional assumption is required.

By substitution into Eq. 37 this becomes

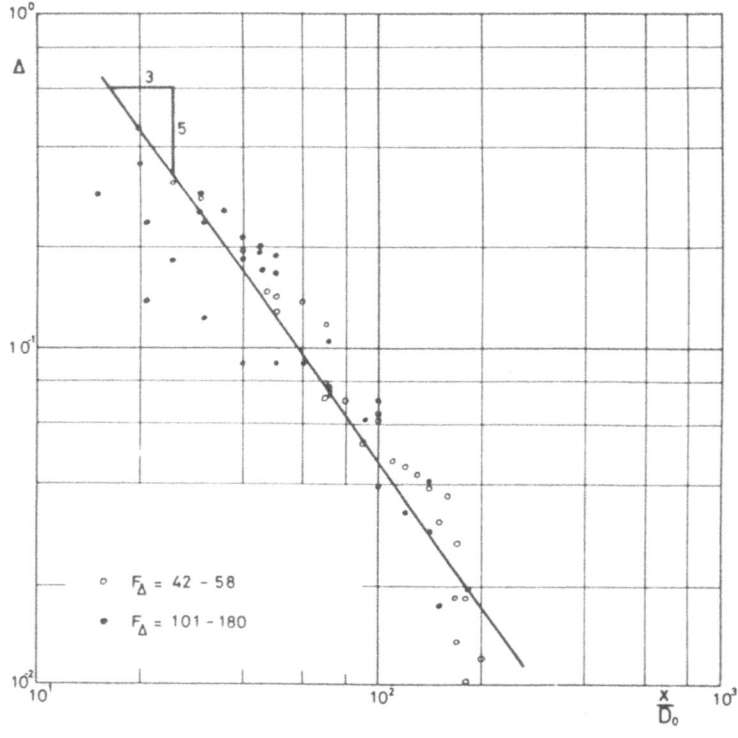

FIG. 5.—COMPARISON BETWEEN THEORETICAL AND EXPERIMENTAL (3) LONGITUDINAL DISTRIBUTION OF DENSITY DEFICIT, MEASURED AT SURFACE CENTER LINE

$$\frac{df_4}{d\xi} = -\frac{28}{9}\,\xi\,f_1^2 \dots\dots\dots\dots\dots\dots\dots\dots\dots \quad (46)$$

To proceed further, the local value of the bulk Richardson number, R_i, is considered:

$$R_i = \frac{\Delta g S}{2\beta v_1^2} = R x_1^{8/3}\,\frac{f_3 f_4}{f_1^2} \dots\dots\dots\dots\dots\dots\dots \quad (47)$$

As the additional assumption, R_i, is taken to be laterally constant, i.e., independent of ξ, this leads to

$$f_4 = f_1^2 \dots\dots\dots\dots\dots\dots\dots\dots\dots\dots\dots\dots\dots \quad (48)$$

When this is combined with Eq. 46, the latter is readily integrated as

$$f_1 = \exp\left(-\frac{7}{9}\,\xi^2\right) \text{ and } f_4 = \exp\left(-\frac{14}{9}\,\xi^2\right) \dots\dots\dots\dots \quad (49)$$

COMPARISON WITH EXPERIMENTS

Even though it may be impossible to make a detailed, direct comparison between theory and published experimental data, at least a crude indication of the validity of the analysis may be obtained by considering the experiments

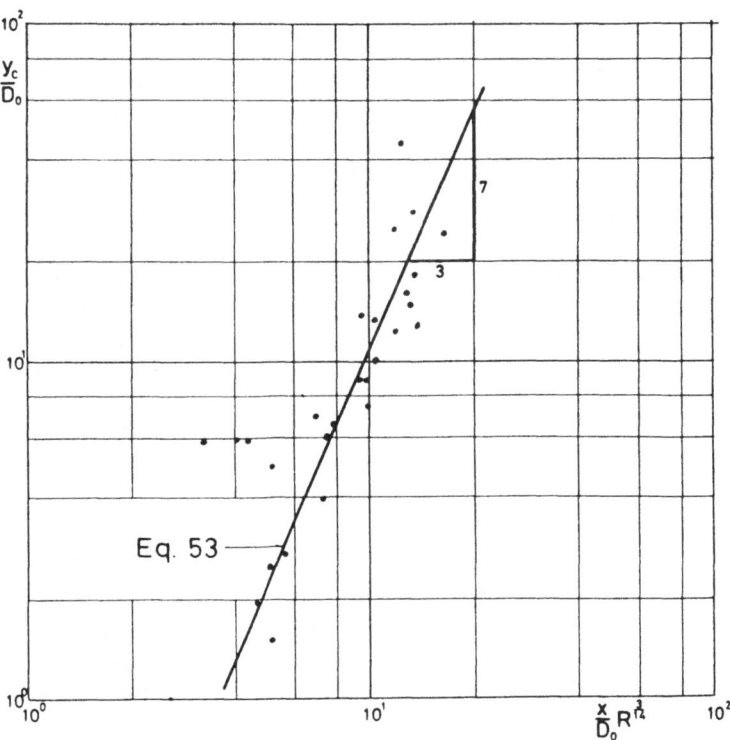

FIG. 6.—COMPARISON BETWEEN EQ. 53 AND EXPERIMENTAL (3) DETERMINATION OF JET SURFACE WIDTH

carried out by Jen and Wiegel (3). These authors considered a horizontal warm-water jet, so that agreement can only be expected if the heat loss to the air is negligible and if the density of the water varies linearly with temperature. The jet was discharged from a pipe with the diameter, D_o, and nothing is reported concerning the extension of the zone of flow establishment. Other difficulties are that the scatter is large, which is natural, and that the velocity profiles along verticals are unknown. Nevertheless, these experiments are considered by the writers to be by far the best for the present purpose.

It has been mentioned previously that the experiments confirm the predicted linear increase of the jet thickness with the distance from the outlet. That the surface velocity distribution in a cross section is Gaussian, as predicted by Eq. 49, is confirmed by Fig. 2 in Ref. 3.

The corresponding distribution of the relative density variation in a cross section, Eq. 49, may be compared with experiments if $\Delta \rho_o / \rho_o$ is substituted for $\Delta T / T_w$, where T_w is the temperature of the ambient fluid, while ΔT is the difference between the local temperature and T_w. The comparison is given in Fig. 4 and is seen to be rather satisfactory. Value y_c is the distance from the jet axis at which the local value of ΔT equals half the value at the axis.

Similarly, the variation of Δ along the jet axis may be compared with experimental data (see Fig. 5). The distance from the outlet is denoted by x, and D_o is the diameter of the pipe. Even though the scatter is rather large, the predicted inclination - 5/3 seems to be a reasonable best fit.

Finally, the lateral spreading of the surface jet may be tested by comparing measured and calculated values of y_c as a function of x. The calculation is based on Eq. 29.

$$\frac{\Delta}{\Delta_o} = x_1^{-5/3} \exp\left(-\frac{14}{9} \xi^2\right) \quad \dots \dots \dots \dots \dots \dots \dots \dots \dots \quad (50)$$

in which $\xi = x_2/(\sqrt{R} \ x_1^{7/3})$. As a reasonable approximation, x_1 is now substituted by x/L and, according to Eq. 45

$$L = 6.67 \ D \approx 6.67 \ D_o \quad \dots \dots \dots \dots \dots \dots \dots \dots \dots \dots \dots \quad (51)$$

The wanted relation is obtained when

$$L \ x_2 = y_c \quad \text{and} \quad \Delta = \frac{1}{2} \ \Delta_o x_1^{-5/3} \quad \dots \dots \dots \dots \dots \dots \dots \quad (52)$$

are inserted.

Then the following relation is obtained:

$$\frac{y_c}{D_o} = 0.053 \left(R^{3/14} \ \frac{x}{D_o}\right)^{7/3} \quad \dots \dots \dots \dots \dots \dots \dots \dots \dots \quad (53)$$

In Fig. 6 this relation is compared with experimental data for all runs corresponding to a local value of $F_\Delta > 2$. The agreement is acceptable except for the three smallest values of y_c. The disagreement here is probably due to a deviation from the similarity assumptions caused by the zone of flow establishment.

The results presented so far are encouraging except for one interesting feature. Eq. 49 implies that the width of the density profile is smaller than that of the velocity profile, which is contrary to the general perception, which seems to be that the two profiles are approximately of equal lateral extent.

Thus, the additional condition necessary to supplement the degenerated system of equations may be

$$f_1 = f_4 \quad \dots\dots\dots\dots\dots\dots\dots\dots\dots\dots\dots\dots \quad (54)$$

which leads to the result

$$f_1 = f_4 = \frac{1}{1 + \dfrac{14}{9}\,\xi^2} \quad \dots\dots\dots\dots\dots\dots\dots\dots\dots\dots \quad (55)$$

which is shown with a dotted line in Fig. 4 for comparison. The corresponding expression for the width, y_c, deviates from Eq. 53 only in the factor which now becomes 0.064.

CONCLUSIONS

The result of the investigation is that a solution of the flow equations is possible if the local value of the bulk Richardson number is less than about 0.1 and if similarity is assumed for the velocity profiles and for the density profiles, respectively.

The solution indicates that:

1. The thickness of the jet varies linearly with the distance from the outlet. The factor of proportionality depends on the entrainment factor and the velocity distribution in a vertical.

2. Both the velocity and the density variation in the lateral direction may be predicted from the theory, Eqs. 49 and 55.

3. A prediction of the lateral extent of the surface jet is made, Eq. 53.

These conclusions are found to be reasonably well supported by experiments.

APPENDIX I.—REFERENCES

1. Ellison, T. H., and Turner, J. S., "Turbulent Entrainment in Stratified Flows," *Journal of Fluid Mechanics*, Vol. 6, 1959, pp. 423-448.
2. Hayashi, T., and Shuto, N., "Diffusion of Warm Water Jets Discharged Horizontally at the Water Surface," *Proceedings of the 12th Congress of the International Association for Hydraulics Research*, Vol. 4, Part 1, Sept., 1967, pp. 47-59.
3. Jen, Y., and Wiegel, R. L., "Surface Discharge of Horizontal Warm-Water Jet," *Journal of the Power Division*, ASCE, Vol. 92, PO 2, Proc. Paper 4801, Apr., 1966, pp. 1-29.
4. Larsen, J., and Sorensen, T., "Buoyancy Spread of Waste Water in Coastal Regions," *Proceedings of the 11th Conference on Coastal Engineering*, London, England, Vol. 2. Part 4, Sept., 1968, pp. 1397-1402.
5. Pedersen, F. B., "Gradually Varying Two-Layer Stratified Flow," *Journal of the Hydraulics Division*, ASCE, Vol. 98, HY 1, Proc. Paper 8679, Jan., 1972, pp. 257-268.
6. Wilkinson, D. L., "Studies in Density Stratified Flows," *Report No. 118*, Water Research Lab., Univ. of New South Wales, New South Wales, Australia, Apr., 1970.

APPENDIX II.—NOTATION

The following symbols are used in this paper:

a, b, c, d, m = ordinary numbers;

B = nondimensional width of surface jet;

D = vertical length scale;

D_o = pipe diameter;

E = entrainment velocity;

e = entrainment factor;

F_Δ = densimetric Froude number at water surface;

f = entrainment function;

g = acceleration of gravity;

L = horizontal length scale;

R = Richardson number;

S = jet thickness;

s = nondimensional jet thickness;

u_i = nondimensional velocity components;

v_i = velocity components;

v_o = reference velocity;

x_i = horizontal coordinates;

y_c = lateral distance;

α, β = correction factors;

Δ = nondimensional density deficit;

Δ_o = reference Δ;

ξ = nondimensional variable;

ρ = density of jet fluid at water surface; and

ρ_o = density of ambient fluid.

9588 SURFACE JET AT SMALL RICHARDSON NUMBERS

KEY WORDS: Buoyancy; **Density;** Diffusion; Discharge (water); **Entrainment; Hydraulics; Jet mixing flow;** Jets; Two phase flow; **Water pollution**

ABSTRACT: The spreading of a jet discharged horizontally at the surface of an initially quiescent water of larger density is treated analytically. The equations of continuity and of mass conservation are derived as well as a momentum equation for the longitudinal and for the transverse direction. The entrainment of ambient water into the jet is included in the analysis. The four partial differential equations thus obtained are then transformed into ordinary differential equations by assuming similarity of the velocity profiles and of the density profiles, respectively. An exact solution of the equations is obtained for the case of small Richardson numbers. From this the following conclusions are drawn: the jet thickness is proportional to the distance from the outlet. The velocity profiles, as well as the density profiles, follow a Gaussian distribution.

REFERENCE: Engelund, Frank, and Pedersen, Flemming Bo, "Surface Jet at Small Richardson Numbers," *Journal of the Hydraulics Division,* ASCE, Vol. 99, No. HY3, **Proc. Paper 9588**, March, 1973, pp. 405-416

Prog. Rep. 58, pp. 31-40, June 1983
Inst. Hydrodyn. and Hydraulic Engrg.
Tech. Univ. Denmark

3. INTERNAL SEICHES IN A STRATIFIED SILL FJORD

by Jacob Steen Møller and Fl. Bo Pedersen

INTRODUCTION

In an earlier report, ref [3], a fieldwork on the stratified sill fjord Affarlikassaa is described. There we argue that major internal seiches due to a resonance phenomenon can occur in the fjord during the autumn season. Lewis, ref. [2], points towards the possible mixing effects of the internal seiche. Here we shall investigate the dynamics of the internal seiche and evaluate the mixing efficiency due to the seiche.

GOVERNING EQUATIONS

Consideration is given to a two-layer and two-dimensional system. The inner fjord is separated from the sea by a sill. The interface between the two layers is situated below the sill depth. Fig. 1 shows the details of the system. The symbols from Fig. 1 are used below. We apply the Boussinesq approximation and assume the velocity field to be one-dimensional. The mixing

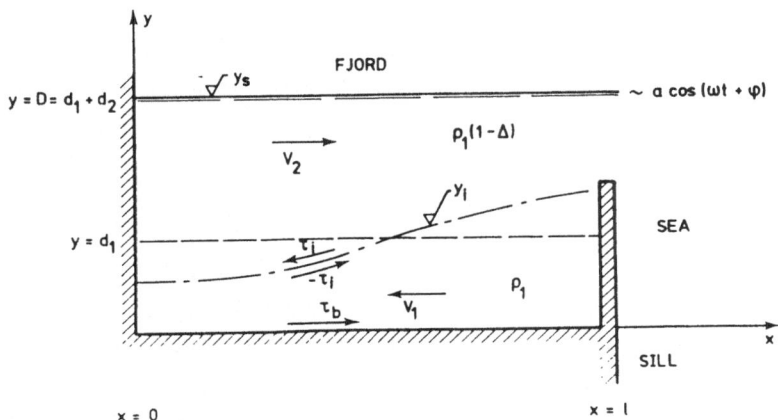

Fig. 1 Definition sketch of the fjord model. The figure is not in scale. For Affarlikassa characteristic values are: $\rho_1 = 1027$ kg/m^3, $d_1 = d_2 = 30$ m, $l = 4000$ m, $a = 0.75$ m and $\omega = 2\pi/T = 1.41 \times 10^{-4} s^{-1}$. For resonance is $\Delta = 2.2 \times 10^{-4}$.

is not taken into account in the analysis of the dynamics.
The equations of continuity are:

$$\frac{\partial}{\partial t} (y_s - y_i) + \frac{\partial}{\partial x} (V_2(y_s-y_i)) = 0 \tag{1}$$

$$\frac{\partial}{\partial t} (y_i) + \frac{\partial}{\partial x} (V_1 y_i) = 0 \tag{2}$$

The equations of motion are:

$$\frac{1}{g} \frac{\partial}{\partial t} (V_2) + \frac{\partial}{\partial x} (y_s + \frac{V_2^2}{2g}) + \frac{\tau_i}{\rho g (y_s-y_i)} = 0 \tag{3}$$

$$\frac{1}{g} \frac{\partial}{\partial t} (V_1) + \frac{\partial}{\partial x} (y_s + \Delta y_i + \frac{V_1^2}{2g}) + \frac{\tau_b - \tau_i}{\rho g \, y_i} = 0 \tag{4}$$

Here τ_i is the friction acting at the interface, and τ_b is the
friction acting at the bottom. The friction is related to the
mean current parameters through the equations:

$$\frac{\tau_b}{\rho} = \frac{f_b}{2} V_1^2 \tag{5}$$

$$\frac{\tau_i}{\rho} = \frac{f_i}{2} \frac{1}{4} (V_2-V_1)^2 \tag{6}$$

where $\frac{f_b}{2}$ and $\frac{f_i}{2}$ are friction factors for the bottom friction
and the interfacial friction respectively. See Bo Pedersen,
ref. [4], for a discussion on the determination of the interfacial friction.

CALCULATION OF THE CURRENTS WITHOUT FRICTION

        Assuming that the friction is negligible, and the ampli-
tudes are small, the Eqs. (1) to (4) can be linearized. A solution to the linearized equations is given by Stigebrandt, ref.
[7]. In the present report we shall seek solutions for slightly
different boundary conditions: When the resonant seiche occurs,
it has a wave length nearly twice as long as the fjord length,
and it has a period equal to the tidal period, T. This suggests
that the seiche is reflecting rather than dissipating at the
fjord head and sill. We represent the tide by a cosine, see

Fig. 1, and by omitting terms of the order of magnitude of Δ we obtain the following periodic solutions for the currents:

$$Q_2 = a\omega l \, A_i \, \frac{d_1}{D} \left[\sin\left(\frac{\omega x}{c}\right) + \frac{x d_2}{l d_1} \sin\left(\frac{\omega l}{c}\right) \right] \sin \omega t \tag{7}$$

$$Q_1 = - a\omega l \, A_i \, \frac{d_1}{D} \left[\sin\left(\frac{\omega x}{c}\right) - \frac{x}{l} \sin\left(\frac{\omega l}{c}\right) \right] \sin \omega t \tag{8}$$

where $Q_1 = V_1 \, d_1$ and $Q_2 = V_2 \, d_2$. c is the internal wave velocity:

$$c = \sqrt{\frac{\Delta g d_1 d_2}{D}} \tag{9}$$

and:

$$A_i = \left[\frac{\omega l c}{g d_2} \cos\left(\frac{\omega l}{c}\right) + \sin\left(\frac{\omega l}{c}\right) \right]^{-1} \tag{10}$$

Resonance occurs when $A_i^{-1} = 0$, which for Affarlikassaa gives $c_{res} = 0.18$ m/s and hence $\Delta = 2.2 \times 10^{-4}$ for the first mode. These values are in good agreement with the observed density differences in the resonance situation, ref. [2].

CALCULATION OF THE CURRENTS INCLUDING FRICTION

Since the maximum speeds and hereby the maximum level of mixing are present at the resonance situation, no detailed knowledge on the mixing can be obtained from the results derived for the frictionless case. We therefore seek a solution to the governing equations including friction.

To reach an analytical solution the problem is divided into the baroclinic case and the barotropic case. The currents for the barotropic and the baroclinic solution are denoted Q_{t1}, Q_{t2} and Q_{c1}, Q_{c2} respectively. The overall solution is found as the sum of the barotropic and the baroclinic currents. The baroclinic case leads to the assumption $Q_{c1} = - Q_{c2}$. Applying this to the governing equations and linearizing them yields after some re-arrangement:

$$\frac{\partial}{\partial t} Q_{c1} + c^2 \frac{\partial}{\partial x} Y_i + \frac{d_2 \tau_b - D\tau_i}{D\rho} = 0 \tag{11}$$

Inserting Eqs. (5) and (6) in Eq. (11) gives

$$c^2 \frac{\partial^2}{\partial x^2} Q_{c1} - \frac{\partial^2}{\partial t^2} Q_{c1} - B \frac{\partial}{\partial t} Q_{c1} = 0 \tag{12}$$

where the friction is linearized by:

$$B = \frac{1}{D^2} \left(\frac{D^4}{4d_1^2 d_2^2} \frac{f_i}{2} + \frac{D}{d_2} \frac{f_b}{2} \right) |Q| = \frac{1}{D^2} \frac{f_T}{2} |Q_k| \tag{13}$$

In Eq. (13) $|Q_k|$ is a constant. The value of $|Q_k|$ is identified as a characteristic "friction discharge". An estimate for $|Q_k|$ is given below. We seek periodic solutions to Eq. (12) and get :

$$Q_{c1} = A_1 e^{\frac{B}{2c}x} \cos\left(\frac{\omega x}{c} + \omega t\right) + B_1 e^{-\frac{B}{2c}x} \cos\left(-\frac{\omega x}{c} + \omega t\right) \tag{14}$$

as the solution for the baroclinic seiche. To find the barotropic current we set $\Delta = 0$ in Eqs. (3) and (4). Since the barotropic discharge at resonance is small in comparison to the baroclinic discharge, Eq. (7), it is resonable to neglect the friction when determining the barotropic current. Doing this we find the solutions:

$$Q_{t2} = a\omega l \frac{d_2}{D} \frac{x}{l} \cos(\omega t + \varphi) \tag{15}$$

$$Q_{t1} = a\omega l \frac{d_1}{D} \frac{x}{l} \cos(\omega t + \varphi) \tag{16}$$

We then add Eq. (14) to Eq. (16) and apply the boundary condition $Q_1 = Q_2 = 0$ for $x = 0$ and $Q_1 = 0$ for $x = l$ and get the solution:

$$Q_1 = a\omega l \frac{d_1}{D} \left(\frac{x}{l} \cos(\varphi) - F \sinh\left(\frac{Bx}{2c}\right) \cos\left(\frac{\omega x}{c}\right)\right) \cos(\omega t)$$

$$- \left(\frac{x}{l} \sin(\varphi) - F \cosh\left(\frac{Bx}{2c}\right) \sin\left(\frac{\omega x}{c}\right)\right) \sin(\omega t) \quad (17)$$

where

$$\tan(\varphi) = \tan\left(\frac{\omega l}{c}\right) \tanh\left(\frac{Bl}{2c}\right)$$

and

$$F = \left\{ \frac{\left(1 + \tan^2\left(\frac{\omega l}{c}\right)\right)\left(1 - \tanh^2\left(\frac{Bl}{2c}\right)\right)}{\left(\tan^2\left(\frac{\omega l}{c}\right) + \tanh^2\left(\frac{Bl}{2c}\right)\right)} \right\}^{\frac{1}{2}} \quad (18)$$

$\frac{\omega l}{c}$ approaches π for the resonance situation, and the barotropic element of Eq. (17) can be neglected. If we use this when seeking the maximum current, $Q_{res,max}$, for $x = \frac{1}{2} l$ we find:

$$Q_{res,max} = a\omega l \frac{d_1}{D} \frac{\cosh\left(\frac{Bl}{4c}\right)}{\sinh\left(\frac{Bl}{2c}\right)} \simeq \frac{2d_1 D a\omega c}{\frac{f_T}{2} |Q_k|} \quad (19)$$

where we have used Eq. (13).

We now want to estimate the magnitude of $|Q_k|$. Our approach is to find the $|Q_k|$ that in the resonance situation and assuming linear friction of the form of Eq. (13) gives the same average τ that we would obtain by calculating the average τ using the quadratic from of Eq. (6). This leads to the criterion:

$$\int_0^1 \int_0^{\pi/2} Q_1^2 \, dt \, dx = |Q_k| \int_0^1 \int_0^{\pi/2} Q_1 dt \, dx \quad (20)$$

We then solve Eq. (20) for $|Q_k|$ and get:

$$|Q_k| \simeq \frac{\pi^2}{8} \frac{d_1}{D} \frac{a\omega c}{B} \quad (21)$$

When we combine Eqs. (21), (19) and (13) we get:

$$Q_{res,max} \simeq \left\{ \frac{2\, ca\omega\, d_1 D}{\frac{f_T}{2}} \right\}^{\frac{1}{2}} \tag{22}$$

The maximum amplitude for the internal seiche, $a_{i,max}$ can now be found to be:

$$a_{i,max} = \frac{T}{2l}\, Q_{res,max} \tag{23}$$

We have plotted the solution for $Q_{res,max}$ for the nonfrictional and the frictional solution against the density difference in Fig. 2. The values used are applicable to Affarlikassaa, and Eq. (21) has been used for all Δ.

DISCUSSION

Fig. 2 shows that the calculated amplitudes of the internal seiche are comparable to the depth of the bottom layer. This indicates that the linear assumption is not satisfactorily fulfilled in the resonance situation. Nonlinear effects are expected for high densimetric Froude numbers, $\mathbb{F}_\Delta > o(1)$. $\mathbb{F}_\Delta$ approaches one in the resonance situation. Further will $\mathbb{F}_\Delta$ increase during the winter due to the decrease in Δ, possibly even the tidal current will produce nonlinear effects. An approach to the nonlinear case is given in ref. [1] for the situation where the interface is located above sill depth.

Topographic effects may also play an important role for the dynamics; such effects have not been included in the model.

THE MIXING EFFICIENCY

The results for the velocity and amplitude of the internal seiche found above are now applied to the theory of entrainment developed in ref. [4]. Since the high velocities and low density differences suggest supercritical flow at peak velocities, we apply a Richardson Bulk Flux number, $\mathbb{R}_f^T$, averaging the subcritical and the supercritical situation. Doing this we find:

$$\mathbb{R}_f^T = \frac{\frac{1}{2}\rho g \Delta\, d_k}{PROD_k}\, V_{ek} = \frac{\frac{1}{2}\rho g \Delta\, d_k}{\int_0^{d_k} \tau \frac{\partial V}{\partial y} dy}\, V_{ek} \tag{24}$$

$$\mathbb{R}_f^T = 0.1$$

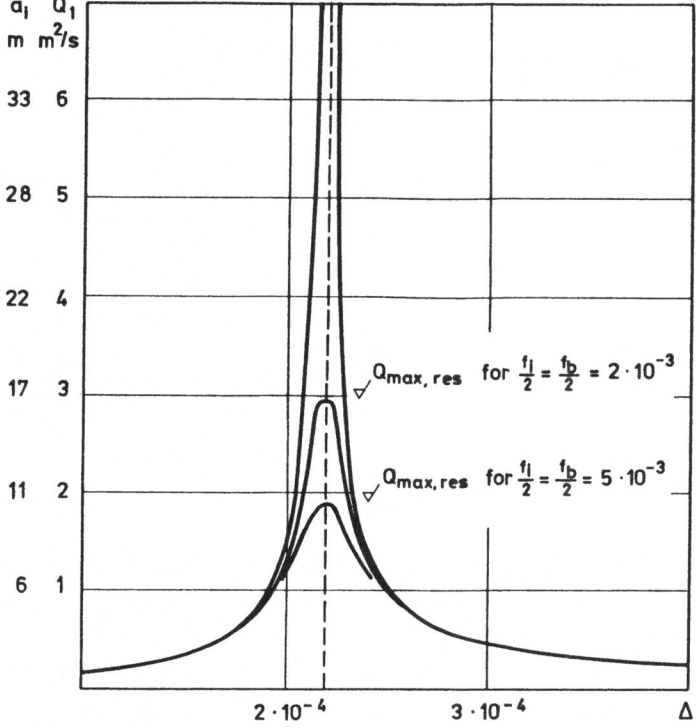

Fig. 2   *Variation of the magnitude of the internal seiche in Affarlikassaa for varying density difference and friction. The used values for the fjord's parameters are mentioned in Fig. 1.*

where V_{ek} is the entrainment velocity, the index k is 1 for the bottom layer and 2 for the surface layer, $PROD_k$ is the production of turbulent kinetic energy for the layer. For a linearly distributed τ the PROD-term is evaluated by:

$$PROD_2 = \tau_i V_2 = \frac{1}{4} \rho \frac{f_i}{2} \left(\frac{D}{d_1 d_2}\right)^2 |Q_k| \frac{1}{d_2} Q_2^2 \qquad (25)$$

where Eq. (6) has been used. The PROD term is integrated over the fjord length and the tidal period giving us the average PROD per area and time given as:

$$\text{PROD}_2 = \frac{1}{2\pi} \rho \frac{f_i}{2} \left(\frac{D}{d_1 d_2}\right)^2 \frac{1}{d_2} \left\{2 d_1 D \frac{ca\omega}{\frac{f_T}{2}}\right\}^{3/2} \tag{26}$$

where Eq. (22) was used. Likewise we obtain:

$$\text{PROD}_1 = \frac{2}{\pi} \rho \frac{f_p}{2} \frac{1}{d_2^{\ 3}} \left\{\frac{2 d_1 D \ ca\omega}{\frac{f_T}{2}}\right\}^{3/2} \tag{27}$$

where the friction factor $\frac{f_p}{2}$ is given by:

$$\frac{f_p}{2} = \left(\frac{f_b}{2} + \frac{1}{4}\left(\frac{D}{d_2}\right)^2 \frac{f_i}{2}\right) \tag{28}$$

We can now use Eq. (24) to evaluate the entrainment velocity into each of the two layers in the resonance case for Affarlikassaa and get: $V_{e,2} = 2 \times 10^{-6}$ m/s for the surface layer, and $V_{e,1} = 4 \times 10^{-6}$ m/s for the bottom layer, assuming $\frac{f_i}{2} = \frac{f_b}{2} = 2 \times 10^{-3}$. If we assume that there is no buoyancy flux across the sill, the change of density for the lower layer due to constant entrainment velocities V_{e1} and V_{e2} is given by:

$$\rho_1(1) \simeq \langle\rho\rangle + (\rho_{1,0} - \langle\rho\rangle) e^{\frac{V_{e1}A}{R_{10}} (1 + \frac{R_{10}}{R_{20}})t} \tag{29}$$

when $V_e At \ll R$.

Here $\langle\rho\rangle$ is the average density for the fjord, ρ_{10} is the initial density for the bottom layer, A is the interfacial areal, assumed constant, and R_{10}, R_{20} are the initial volumes of the two layers.

Observations from Affarlikassaa suggest that the period of resonance occurs over a period less than a week. The change in Δ calculated from (29) using values from Affarlikassaa during resonance is about 10% over one week. This is in good agreement with the features of Fig. 2, since a reduction of Δ of about 10% will reduce the seiching drastically.

CONCLUSION

The present investigation shows that the mixing efficiency of the seiche is small compared to other mixing processes, one major reason for this is that the seiche is "tuned off" by the mixing.

The large amplitudes of the seiche in Affarlikassaa, however, could play an important role in removing bottom water, since amplitudes of about 15 m will cause spill of bottom water across the sill. ISVA's measurements in Affalikassaa show a rapid lowering of the interface depth when resonance conditions are present, Fig. 3. This lowering approximately from 30 m to 40 m, could be the result of such a spill. To determine whether this actually is the case, further investigations have to be made.

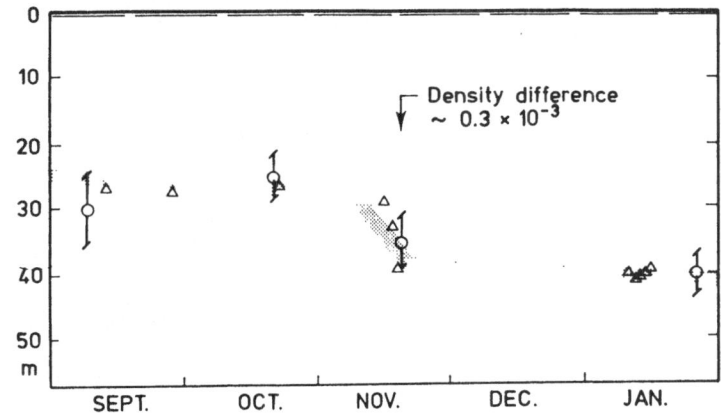

Fig. 3   Location of halocline in Affarlikassaa Fjord, Maarmorilik, Greenland.
Δ: Measurements by ISVA
o: Measurements by Greenex, ref. [5] and [6].

REFERENCES:

[1]   Blackford, B.L., 1976: On the generation of internal waves by tidal flow over a sill - a possible nonlinear mechanism. Jour. of Marine Research. (36,3, 1978). pp. 529-549.

[2]   Lewis, E.L., 1980: Water Movement in A and Q Fjords near Marmorilik and the Processes for Pollutant Transport. Institute of Ocean Sciences, Box 6000, Sidney B.C. Canada. Report to the Ministry of Greenland, Copenhagen.

[3]   Møller, J.S., Bo Pedersen, Fl. and Nielsen, T.K., 1982:
        Measurements of the convective mixed layer under growing
        sea ice. ISVA, Technical Univ. of Denmark, Prog. Rep. 56,
        pp. 13-23.

[4]   Pedersen. Fl. Bo, 1980: A monograph on turbulent entrain-
        ment and friction in two-layer stratified flow. Chapter
        1, 2, 3 and 5. ISVA, Technical Univ. of Denmark, Series
        Paper No. 25. Diss.

[5]   Pedersen, K., 1980 and 1981: Resipient-kontroldata fra
        havvand, A-fjord st. 4 og Q-fjord st. 10, 1980 and 1981,
        (Sea water quality data for A-fjord st. 4 and Q-fjord
        st. 10, 1980 and 1981). Internal report in Danish, Greenex
        A/S, Copenhagen.

[6]   Pedersen, K., Greenex A/S: Personal communication.
        (Data for 1981 - 1982).

[7]   Stigebrandt, A., 1976: Vertical Diffusion Driven by Inter-
        nal Waves in a Sill Fjord. Journal of Physical Oceano-
        graphy, Vol 6., July 1976, pp. 486-495.

Prog. Rep. 54, pp. 33-40, August 1981
Inst. Hydrodyn. and Hydraulic Engrg.
Tech. Univ. Denmark

ON ARCTIC LAKES

A THERMODYNAMIC AND HYDRODYNAMIC INVESTIGATION

by Niels Danielsen, Steffen K. Iversen and Fl. Bo Pedersen

INTRODUCTION

Hydropower plants in the arctic regions are typically of
the high-head type. The meltwater runoff takes place during a
few months in the summertime and are normally stored in a
natural lake, which, with only minor modifications, can be used
as a magasin for the hydropower plant. The temperature of the
meltwater is 0-1°C, depending primarily on the head difference
between the lake and the source. During the summertime the lake
water is warmed and may/may not exceed the temperature of maximum density at 4°C. In the actual lake, located at Marmorilik,
Greenland (70°40' north), the lake water temperature keeps below
4°C. As can be recognized by the air-temperature diagram in
Fig. 1, permafrost in the soil/rock is present at this location.
Hence ice-formation in the head-race tunnel is a possibility
during the start-up period. Later on a sudden temperature rise
in the water or a water-hammer may cause an ice-gang in the
tunnel, which will block the tunnel substantially. A thorough
investigation of the ice-problems and the thawing-problems in a
permafrost embedded water-tunnel calls for a detailed knowledge
of the thermo- and hydrodynamics of the reservoir from which
the water is withdrawn.

For the Upper South Lake at Marmorilik we have one year
of rather intense meteorological observations, which have been
used to interconnect the very sparse temperature and ice-
thickness observations in the lake and to calibrate the empirical formulae for the various heat-exchange-terms. Normally,
time-series for the air temperature are the only meteorological
parameter available in Greenland, and hence we have investigated
the possibility of making a heat-budget based on the temperature
time series and some average values of the other meteorological
parameters.

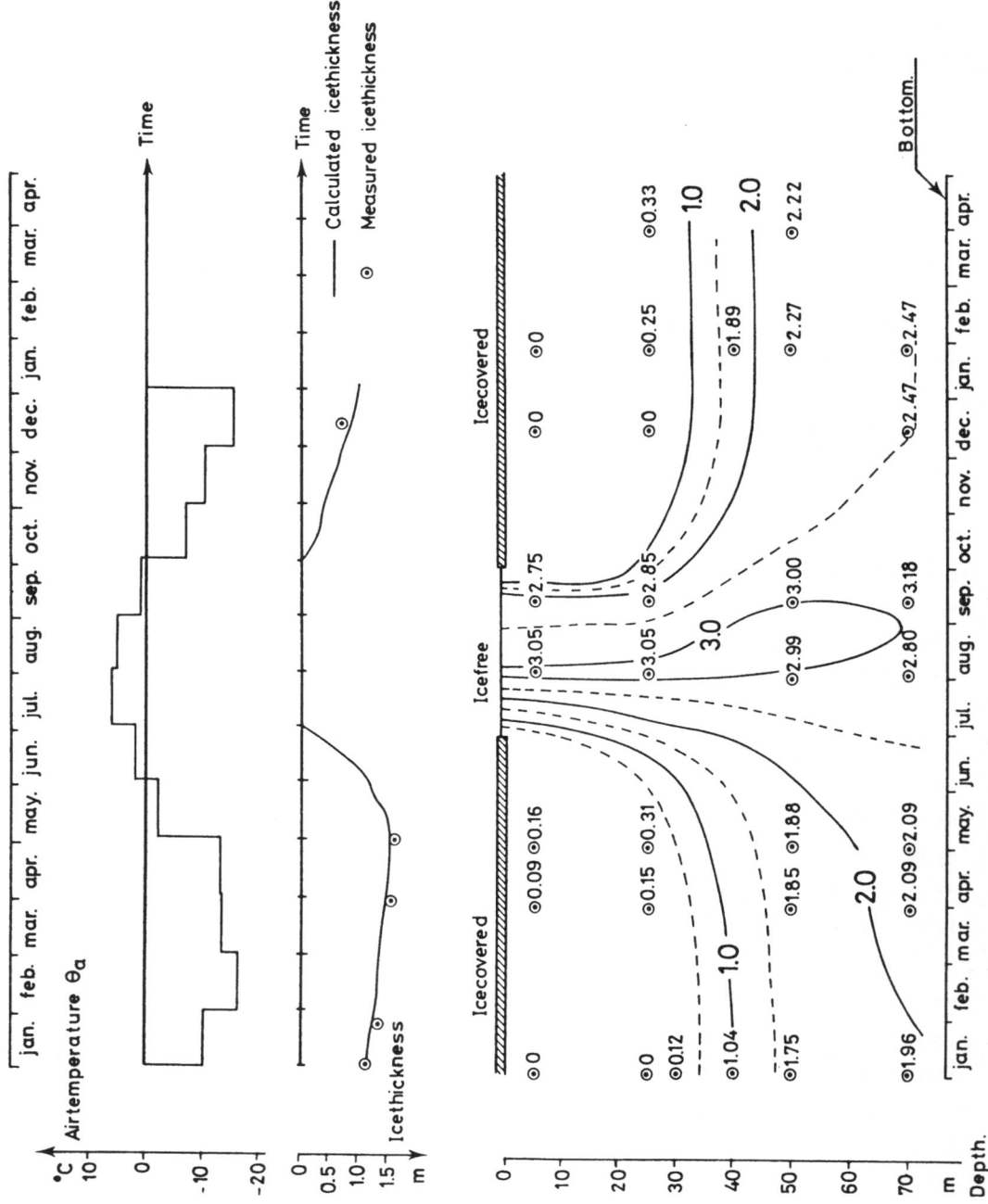

Fig. 1 Isotherms in Upper South Lake 1980/81
Numbers: °C

1. THE SEASONAL VARIATION

In Fig. 1 the annual variation of the water temperature in the lake, the air temperature and the ice thickness is shown. In the ice-covered period, the lake is temperature stratified, but it will be homogenized after only a couple of weeks exposed to the wind and the sun. The wind and the free penetrative convection (due to heat) both play a part in this process of homogenization. The temperature stratification will rebuild in the early autumn when a heat loss from the water surface is present. The thermocline formation takes place in the very short period of the ice-free conditions, and hence the wind during September/October plays a crucial role for the temperature stratification present in the succeeding winter. The annual cycle in temperature is opposite to the one known from lakes in more temperate regions, because the water temperature keeps below $4^{\circ}C$.

2. THE ICE-FREE PERIOD

2.1 CALCULATION OF THE ENERGY BALANCE

During the ice-free period the essential contributions to the energy changes between the lake and its surroundings are:

Net. shortwave radiation:
$$R_{nK} = I_0 \, a \, (1-r) A$$

Net. longwave radiation:
$$R_{nL} = -5.5 \times 10^{-8} (\theta_0 + 273)^4 (1 - \alpha - \beta \sqrt{\overline{e_a}})(1 - 0.8c)A$$

Convective energy transport:
$$R_h = \rho_a \, c_a (\theta_a - \theta_0) c_h \, W \, A$$

Energy exchange by evaporation/condensation:
$$R_e = L \rho_a \frac{0.622}{P} (e_a - e_{0,m}) c_e \, W \, A$$

Energy exchange due to fresh water in- and outflow

in: $R_i = \rho_w c_w \theta_i Q_i$

out: $R_u = \rho_w c_w \theta_u Q_u$

where

I_0	= daily mean insolation	in W/m^2
r	= albedo	dimensionless
a	= transmission factor	dimensionless

A	= area of the lake	m^2
$\theta_i, \theta_u, \theta_0, \theta_a$	= temperature	$^\circ C$
$e_a, e_{0,m}$	= vapor pressure	mm Hg
c	= relative cloudiness	dimensionless
ρ_a, ρ_w	= density	kg/m^3
c_a, c_w	= heat capacity	$J/kg\ ^\circ C$
$c_{e,h}$	= empirical constants	dimensionless
W	= wind velocity	m/s
P	= absolute airpressure	mm Hg
L	= latent heat of fusion	J/kg
Q_i, Q_u	= streamflow	m^3/s
α, β	= empirical constants	dimensionless

Following indices are used:

0	= surface
0,m	= surface, saturated
a	= air
w	= water
i	= inlet
u	= outlet

Using these parameters, the energy balance will be in watt. The equations for R_{nK}, R_{nL}, R_e and R_h are empirical and shall therefore be calibrated against the measured energy balance for the lake.

2.2 CALIBRATION

The equations for R_e and R_h are based on the Reynolds' analogy. Hence the two coefficients c_e and c_h have been taken equal, which implies that the total energy balance becomes rather insensitive to these coefficients, simply because the sum of R_e and R_h is small.

In the equation for net shortwave energy supply R_{nK} a transmission coefficient a is used, which depends on the relative cloudiness c. A linear relationship between a and c has been used within the limits of a(no clouds) = 0.73 and a(totally clouded) = 0.18.

The net longwave radiation R_{nL} can a priori be estimated relatively large due to the arid climate in the area $(e_{a,max}$ = 6 mm Hg). The calibrated constants α = 0.30 and β = 0.042 confirm this (Snow Hydrology, 1956).

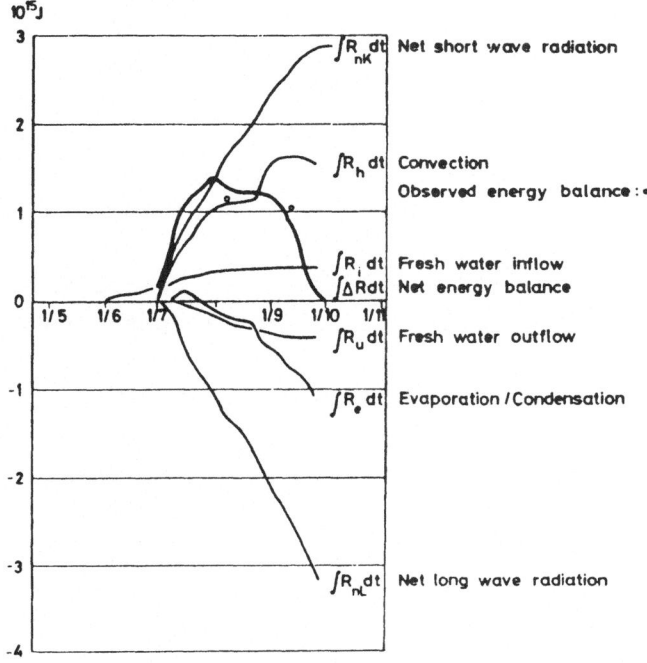

Fig. 2 The calculated and calibrated energy balance for Upper South Lake, 1980 with two observed values plotted.

The basic meteorologic and hydrologic data are wind speed, surface temperature, streamflow and relative cloudiness, which have been measured intermittently at Upper South Lake. In the periods with lack of data these values as well as data for the air temperature, the absolute air pressure and the vapor pressure in the atmosphere are estimated by use of data from Marmorilik Heliport, 10 km from the lake.

Fig. 2 shows the calculated and calibrated energy balance when all data available are used.

2.3 SIMPLIFIED MODEL FOR CALCULATION OF THE ENERGY BALANCE

Often the only available meteorologic data in Greenland are a temperature time series. Therefore, we have tried to estimate the energy exchange budget for the lake with the known temperature time series as the only varying parameter. Some comments on the other contributions are pertinent:

The sum of the two *advective contributions* R_i and R_u is so small that both may be neglected in this connection.

When calculating the *net shortwave radiation* it is necessary to know the relative cloudiness c, the insolation I_0 and the albedo r of the water surface. I_0 can be calculated when the geographic location of the lake is known. r is quite well known as a function of the altitude of the sun. If a mean value for c is used, one can expect that this value is representative for even shorter periods.

When calculating the *net longwave radiation* it is necessary to know the absolute surface temperature $T_0 = \theta_0 + 273 [K]$, the vapor pressure of the air and the relative cloudiness. The absolute surface temperature is quite well known. The radiation is rather insensitive to a variation in the vapor pressure c_a.

The *convective energy transport* can be calculated with knowledge of the difference in air- and surface temperature and the wind speed. Hence, a qualified guess for the summer average wind speed is needed. The water-surface temperature will lie in the narrow range from $0°C$ to $4°C$.

When calculating the energy exchange by *evaporation/ condensation* it is necessary to know the absolute air pressure, the saturated vapor pressure at the surface and the wind velocity. As air temperature is the only available parameter, vapor pressure will be determined by use of this temperature under the assumption that the relative humidity is constant.

Fig. 3 shows the energy balance calculated by use of all available data as well as by use of the air temperature as the only variable.

The conclusion is that the correct order of magnitude of the energy exchange in an arctic lake can be calculated just by taking the varying air temperature into consideration. The vapor pressure of the air must be considered to be the second most important parameter, next to the temperature.

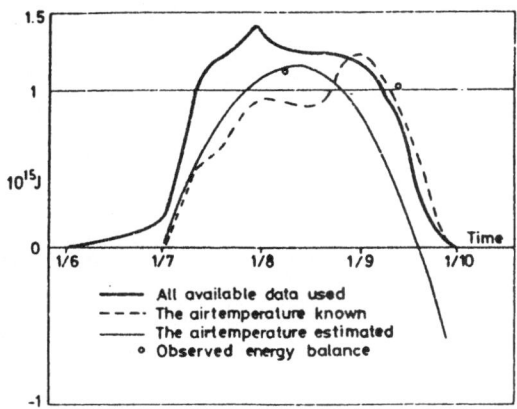

Fig. 3 The energy balance based on three different methods of calculation.

3. THE ICE-COVERED PERIOD

The duration of the ice-covered period has a large bearing on the thermodynamics of the lake and was therefore calculated as well.

The growth and decay of an ice-cover depends on the same elements as the energy balance. A thermodynamic model was developed, and the annual variation of the ice-thickness simulated as shown in Fig. 1.

If the air temperature is the only known parameter, one can use the model of Stefan (1891):

$$h(t) = \kappa_1 \sqrt{K(t)}$$

where h is the ice thickness in m, κ_1 an empirical constant in m (deg.day)$^{-\frac{1}{2}}$ and K(t) the sum of degree days defined as:

$$K(t) = - \int_0^t \theta_a \, dt$$

where θ_a is the air temperature in $^\circ$C.

The empirical constant κ_1 is by calibration (see Fig. 4) determined to 0.025 m/(deg.day)$^{\frac{1}{2}}$, equal to 70% of the theoretical value, which does not take net shortwave radiation and snow cover into account:

$$\kappa_{1,\text{theoretical}} = \sqrt{\frac{2\lambda}{\rho c_p}} = 0.035 \text{ m/(deg.day)}^{\frac{1}{2}}$$

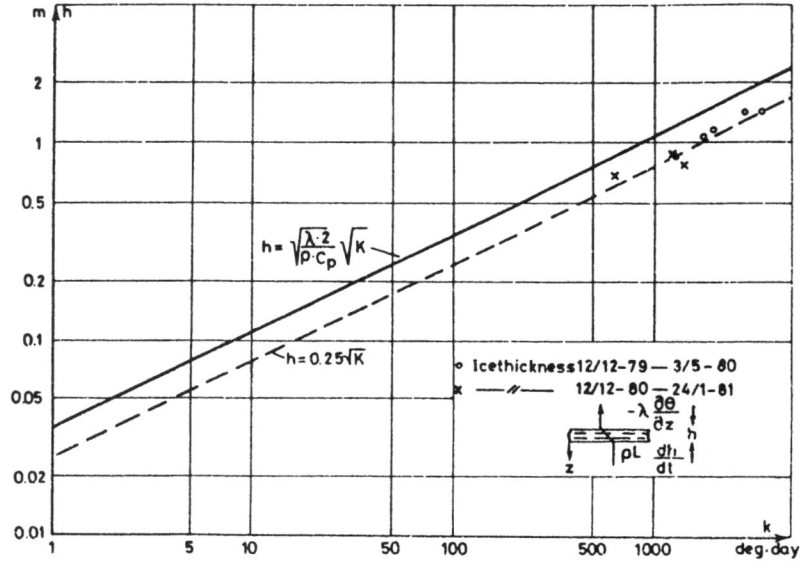

*Fig. 4 Calculated and measured ice-thickness
in Upper South Lake.*

Stefan's model is only intended to simulate the growth of
the ice. During the melting-season one can similarly as a first
approximation use a linear relationship between the ice-thickness
h[m] and the sum of degree days K, which is

$$h(t) = h(0) + \kappa_2 K$$

the empirical constant was found to be $\kappa_2 = 0.014$ m/deg.day.

4. CONCLUSION

In the Upper South Lake the sensitive energy budget is dominated by the energy exchange with the surroundings during the
period when the lake is ice-free. It is possible to give a
reasonable estimate of the energy exchanges and the duration
of the ice-free period, even if the air temperature is the only
known parameter.

References:

Arktiske søer: Steffen Kammeyer Iversen and Niels Kock
Danielsen; advisor Fl. Bo Pedersen. ISVA June 1981.
Master thesis (in Danish).

Prog. Rep. 54, pp. 35-45, sept.1984
Inst. Hydrodyn. and Hydraulic Engrg.
Tech. Univ. Denmark

3. A MULTIPURPOSE STRATIFIED FLOW FLUME

by Flemming Bo Pedersen

INTRODUCTION

The growing environmental concern calls especially for an improved understanding of the basic physics of stratified flows in oceanography, meteorology as well as in civil engineering. The wide range of spatial and time scales encountered in nature combined with a wide spectrum of simultaneously imposed forcing functions - such as wind, tide, runoff, heating/cooling/freez-ing - makes it difficult and uncertain to interpret the in situ measurements. Hence, measurements under controlled laboratory conditions are advantageous, but as we intend to model complex nature, a high degree of flexibility in geometry and in imposed forcing functions is needed in the laboratory set-up. At our institute a stratified flow flume has been developed which to a great extent fulfils this demand for flexibility. First the layout and the basic ideas of the flume are presented. Next, a number of performed experiments within stratified flows are addressed. The main objective of the article is to promote the highly needed research on stratified flow by illustrating how to build and run a relatively inexpensive, multipurpose stra-tified flow flume.

THE STRATIFIED FLOW FLUME

The geometry

The frame in the experimental set-up is a simple 2 meter long tank with rectangular cross section of width W and depth D, preferably with glass walls. The flexibility in geometry is obtained by adjusting the elements inserted in the tank. An artificial bottom and lid, consisting of permeable/impermeable

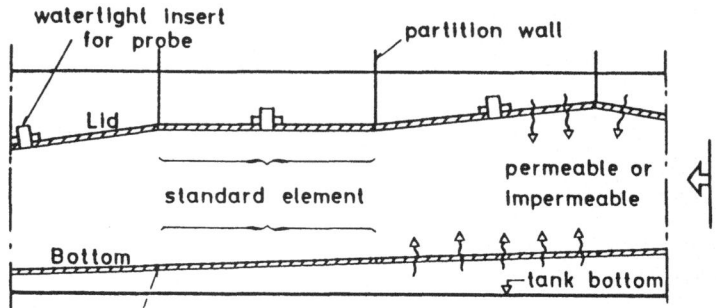

Sealed by air-pressure tubing along circumference
of standard element.

*Fig. 1   Sketch of tank section with inserted upper and
lower vertical standard elements.*

standard elements, can be installed in the tank to form a
linear or polygonal upper and lower boundary for the flow, see
Fig. 1.

The dimensions of the flume can be chosen in accordance
with existing tank/flume facilities, although some optimum dimensions can be outlined, depending on the phenomena to be investigated. Table 1 gives the dimensions of our facilities
(existing and contemplated), and Fig. 2 shows an experimental
set-up to demonstrate a light roof current with entrainment
(from below) and with boundary supply of stable or unstable
water (from above). A brief discussion of the optimum dimensions
is given below.

Purpose	Length (m)	Width (m)	Depth (m)
Demonstration flume (built)	2.95	0.15	0.49
Research flume (contemplated)	10-20	0.20	1.0

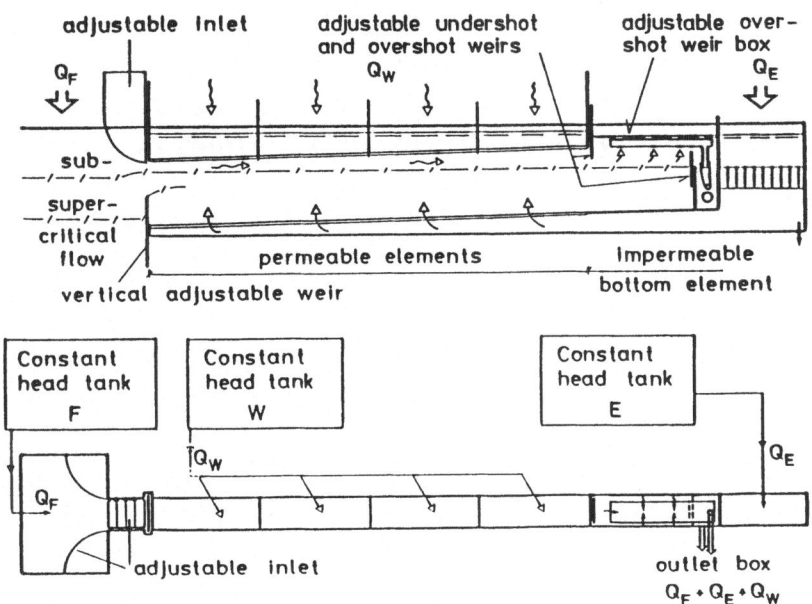

Dimensions of test section: L = 2.95 m; W = 0.15 m;

H = 0.49 m.

Fig. 2   The demonstration flume. Experimental set-up for a light roof current (fresh water discharge Q_F) with entrainment of salt water (discharge Q_E, salinity S_E) and with a forced water supply through the ceiling (discharge Q_W, salinity $S_W \lessgtr S_E$).

The water supply

A constant (or variable) rate of flow can be supplied to the flume from three different constant head tanks, see Fig. 2. The discharges are:

Q_F: This is normally the primary water supply, and hence the inlet consists of a separate inlet box with guiding walls, first vertical then horizontal. The water supply will typically be constant temperature fresh water, but heated water/ salt water can be used as well.

$\underline{Q_E}$: The bottom water supply can either be a main or a secondary
source. An example of the first type is a two-layered paral-
lel current or counter current, and an example of Q_E as the
secondary source is entrainment into a light roof current.
The Q_E supply will typically be salt water, but temperature
differentials can also be used. Q_E can be varied in space
and time (see below).

$\underline{Q_W}$: The imposed discharge through the lid in the tanks can be
a dynamically primary supply, as for instance in experi-
ments with free penetrative convection (unstable) or "katabatic winds" (stable) or it can be a secondary supply with
a more modest dynamic effect on the flow. Depending on the
experiment, Q_W can be heated, cooled, fresh, salt etc. Q_W
can be varied in space and time (see below).

Q - variation

A time variation of all the discharges can be achieved by
governing a pump, a valve, the pressure head or similar. The discharges Q_E and Q_W can be unevenly distributed along the flume
by selecting the filter-characteristics of the elements, and
Q_W by imposing different discharges at each element section,
which is separated by a watertight partition plate. In Fig. 3
the element construction is shown. It consists of a rigid rectangular frame, which supports the air-tight sealing tube along
the periphery and provides a frame for the two stainless steel
perforated cover plates, between which the interchangeable filter
is held. The filter serves *two* purposes. Firstly, it ensures an
even distribution of the flow in an element. Secondly, it prevents an upward flow through the lid element when a test with
unstable conditions is run (water "W" denser than water "F").
Condition two is normally the most restrictive for the filter
design. The filter is an 8 mm thickness of cellulose and in-
organic filtration aids. It has a characteristic of laminar
flow, i.e.

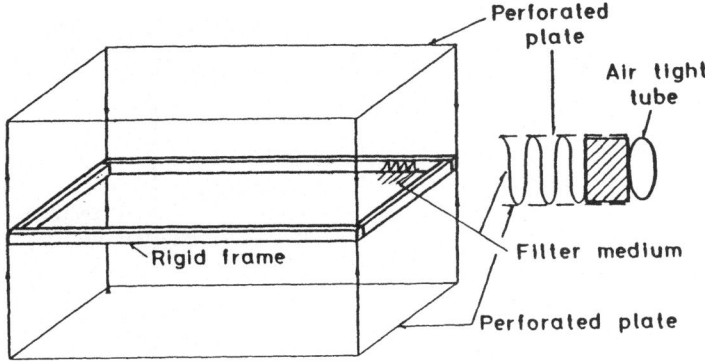

Fig. 3   Standard element in the demonstration flume.

$$\Delta H[mm] = 83\ V_W\ [mm/s]\quad ^{*)}$$

where ΔH is the pressure drop and V_W is the filter velocity.

Q - identification

Besides knowing the basic discharges Q_F, Q_E and Q_W, we may be interested in identifying their respective contribution to the flow in any one cross-section, see Fig. 4. As the buoyancy agent for the upper (W) and the lower (E) water supply often is identical (salt or heat) and because the concentration in the main flow typically is very low, it is advantageous to use different tracers for the upper, the lower and the main flow, respectively. Depending on the available instrumentation, different colours, radioactive tracers or similar can be used. In this way the tracer concentration distribution can be measured with high accuracy, which in turn gives an accurate density distribution. Combined with a detailed velocity (and also preferably turbulence) measurements, the mixing processes under various stable/unstable conditions can be investigated very accurately.

---

*)   Comment: When salt water is used, it is necessary to pre-
     filter the water in order to avoid clogging and
     hence changing the filter characteristic.

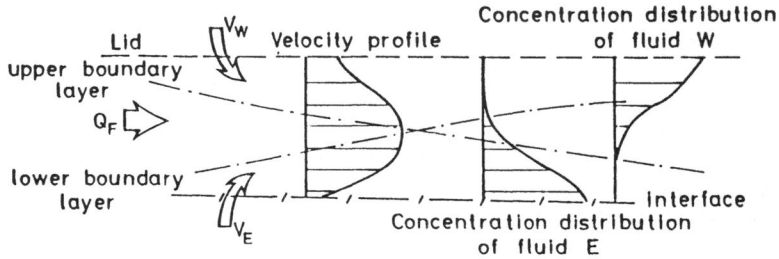

Fig. 4 Sketch of lower and upper boundary layer.

SCOPE OF EXPERIMENTS

The flume is primarily designed for the investigation of
density stratified flows of a complex nature, i.e. flows with
more than one imposed forcing function or flows with a complex
boundary configuration. Much knowledge has been gained during
the last decades on the dynamics of individually imposed forcing types of density stratified flows, such as jets, dense
bottom currents, stratified lake exposed to the wind, free penetrative convection, to mention but a few, while for instance
a jet in a cross-current or a stratified lake simultaneously
exposed to the wind and to free penetrative convection still
contains many unsolved problems, especially concerning the mixing across the interface. Similarly, the damping of the turbulence due to density stratification is a great unsolved problem
within stratified flows, even in the simple density stratified
flows.

Unfortunately, laboratory experiments with density stratified flows are expensive either in heat or in salt consumption.
Therefore, it will be necessary carefully to analyse the necessary flow conditions including the width and the length of the
flume with regard to the phenomena to be studied. Basically, the
following two types of experiments may be of interest, namely
generating turbulence, internally (at the interface) or externally (at the walls). The phenomena to be studied and the cost
to establish and run these two types of experiments are quite
different, as elucidated below.

Internally generated turbulence

Generally, experiments that focus on the phenomena governed
by interfacially generated turbulence are the most expensive.
First, because of the lower friction at the interface compared
to a fixed wall (a ratio of say 1:4) the width to depth ratio
should be at least 10 - 15 in order to compensate for the side
wall effects in a density stratified flow. Secondly, the damping of the eddy viscosity for heat/salt in the equilibrium layer
(the interfacial region) produces a damped rate of growth of
the heat/salt boundary layer, which means that the length of the
flume has to be extended to say 500 times the depth in order to
get a fully developed density and hence a fully developed velocity profile - compared to a length of say 50 times the depth in
ordinary open channel flow. Thirdly, the instrumentation and
the Reynolds' number put a lower limit to the depth and to the
discharge per unit width, respectively. In summary, an experimental facility for basic research on phenomena governed by
internally generated turbulence in density stratified flows is
too expensive to build and run for most laboratories - including ours, for the time being. Therefore, we have turned to the
buoyancy flows, where the turbulence - and hence the mixing -
is dominated by externally generated turbulence (at the fixed
walls).

Externally generated turbulence

If we accept wall generated turbulence as the primary mix--
ing agent, there is then no serious restriction on the width
to depth ratio in the flume, whereby the heat or salt consumption is considerably reduced. Similarly, the higher level of
turbulence created by the fixed walls compared to that generated
at the interface, reduces the length to depth ratio. Hence, the
initial cost of the flume and the running expenses become reasonable.

The major design criteria for our new stratified flow flume
was that it should be inexpensive, flexible, easy to run and -
most important -- able to create buoyancy flows with two or more
simultaneously imposed forcing functions. Furthermore, the basic

knowledge of the flow imposed to the one or the other forcing
function acting separately should be well-known. The choice fell
on: light roof current (a reflected image of dense bottom current) and free penetrative convection. In the following we shall
elaborate on this choice.

Light roof current (dense bottom current)

A light roof current or a dense bottom current is the flow
created by a source of mass, momentum and buoyancy flowing into
an ambient fluid in such a way that the flow is bounded by the
fixed walls and the interface. They play an important role in
the oceans (for instance : The Denmark Strait bottom current,
discharge 5 - 10 × 10^6 m^3/s), in the estuaries (for instance:
spill over the sill in fjords), in the lakes and reservoirs
(river entering with higher/lower density), in the meteorology
(dust storms, katabatic winds, avalanches), in the mining indu-
stry (methane flow in tunnels, tailings outlet etc.) Consequently, from a practical point of view, the knowledge of the dynamic behaviour including the mixing of this type of density currents is important. This need has enhanced the scientific effort
on the subject. Research on mixing problems in dense bottom
currents was initiated by Ellison and Turner [1959], followed
by Löfquist [1960] and later by a great number of scientists,
see Bo Pedersen [1980a, b] for references. Hence, our present
knowledge of the physical processes occurring in light roof
currents or dense bottom currents is fairly good, which gives
this type of current an obvious preference in the choice of a
basic flow in a research on more complex flow-situations. The
light roof current has been preferred to the dense bottom current for economic reasons, as the light roof current can be run
with fresh water as the basic discharge, which means that only
the smaller compensating entrainment discharge from the ambient
fluid is the more expensive salt water. The only serious dis-
advantage is the lid, which puts some restriction on the probe
position.

Free penetrative convection

Free penetrative convection is the turbulent movement without a mean velocity created by a source of buoyancy flux flowing into an ambient fluid. In geophysics, the free convection
often stems from the heating or cooling of water/air. In the
oceans, the estuaries and in the lakes free convection is one
of the major geophysical processes responsible for the seasonal
variation of the thermocline. In the arctic region the salt ex-
clusion from a growing sheet of sea ice causes free penetrative
convection which may have a major effect on the dynamics of the
water body below, and similar effects can be encountered in the
region due to evaporation. In meteorology, the stratification -
stable at night - is often successively destroyed during the daylight hours by free penetrative convection created by the positive heat flux from the ground. It is typical of most of the
penetrative convection cases in nature that they are at the
same time combined with other turbulence generating mechanicms,
for instance wind. Hence, nature gives ample motivation parti-
cularly for making research on free convection combined with
another type of flow. Furthermore, our basic knowledge of the
"pure" free penetrative convection is fairly good, especially
thanks to the meteorologists who have been dealing with the problem since the early forties, while the oceanographers have been
more modest in their contribution until recently. A historical
summary can be found in Bo Pedersen [1980], where the physics
and the governing equations are also given. There is also an excellent review paper by Tennekes and Driedonks [1980].

Examples of experiments

It is to be emphasized that the dimensions of the demonstration flume generally do not meet the requirements for scientific
experiments - the demonstration flume is merely an easy, inexpensive tool for pilot tests. The most severe restriction of
the demonstration flume is its short length.

The flume has been applied for a year now. The first series
of experiments was primarily devoted to testing and modifying

the flume. Light roof-currents with and without imposed free
convection were investigated.

In the second series of experiments, which was performed by
Møller [1984], buoyancy driven circulation in a fjord (salt re-
jection from a growing sheet of sea ice) or in a lake (heat
loss from cooling water) was investigated, see Fig. 5.

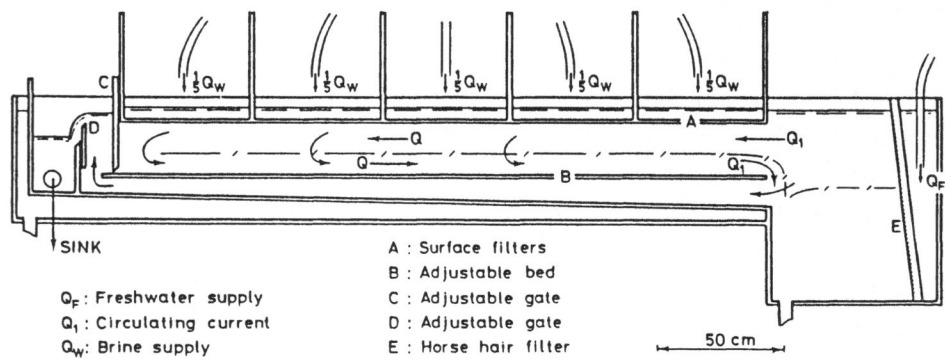

Fig. 5 Experimental set-up for buoyancy driven circulation.

Brine was used to simulate the salt-rejection/heat loss (Q_W).
The density and velocity profiles were measured along the fjord-
arm and compared with the theory and so was the induced circulation discharge Q_C. The experiments and the theory are reported
on in full in the doctoral thesis by Møller [1984] and favour-
ably compared with field measurements from an Arctic fjord.

In the third series of experiments performed, special attention was paid to the two basic currents. The light roof-currents
were – as expected – greatly influenced by the small flume
length, which prohibits the boundary layer of the entrained
salt water (Q_E) to develop fully. The free penetrative convection experiments did not have this drawback, and hence the results are of scientific interest, inasmuch as they cover an entrainment range (very low values), which has not been reported
on in the literature before, and which are in a range encountered in the field. The experiments are reported in Jürgensen
and Pedersen [1984].

REFERENCES

[1] Bo Pedersen, Fl. [1980a]. "A monograph on turbulent entrainment and friction in two-layer stratified flow". Series Paper 25, Institute of Hydrodynamics and Hydraulic Engineering, Technical University of Denmark, 1980.

[2] Bo Pedersen, Fl. [1980b]. "Dense bottom currents in rotating ocean". Journal of the Hydraulics Division, Proc. of the American Society of Civil Engineers, Vol. 106, No. HY 8, Aug. 1980.

[3] Ellison, T.H. and Turner, J.S. [1959]. "Turbulent entrainment in stratified flows". Journal of Fluid Mechanics, Vol. 6, 423-448.

[4] Bo Pedersen, Fl. and Jürgensen, C.[1984]. "Laboratory experiments on entrainment due to free convection". Progress Rep. No. 61, Institute of Hydrodynamics and Hydraulic Engineering, Technical University of Denmark.

[5] Löfquist, K [1960]. "Flow and stress near an interface between stratified liquids". Physics of Fluids, Vol. 3, 158-175.

[6] Møller, J.S. [1984]. "Hydrodynamics of an Arctic fjord". Series Paper 34, Institute of Hydrodynamics and Hydraulic Engineering, Technical University of Denmark.

[7] Tennekes, H. and Driedonks, AGM [1980]. "Basic entrainment equations for the atmospheric boundary layer". Second IAHR Symposium on stratified flows, Trondheim 1980. Editors: T. Carstens, Th. Mc Climans.

SUBJECT INDEX

A

Acceleration of gravity,
  reduced, 8, 17, 160
Alberni Inlet, 129-131
Arctic lakes, III.21, 1-8
Atmospheric inversion
  rise, 99-100

B

Baltic Sea, 102-105, 116-117,
      III.18, 1-20
Baroclin circulation, III.20,3-5
Barotrop circulation, III.20,3-5
Bernouilli's eqt., 17, 24-25
Bornholm Basin, 116-117
Boussinesq approximation, 18
Bruunt-Vaiasäla frequency, 138
Buoyancy driven circulation,
  100-102
Buoyancy flux
  by diffusion, 61
  by entrainment, 61, 107-108
  by evaporation, 14-16
  by heating/cooling, 102-105
  by ice growth, 100
  source of, 99-112

C

Cabbaling, 38
Colebrook and White, 72
Compressible fluid, 37-39
Consequence analysis, III.18,1-20
Continuity eqt., 37-41
  for dense bottom current,
      41, 97
  for wind driven erosion, 123
  for wind mixed Fjord, 125
Convection, see Free penetrative
  conv.
Coriolis force, III.15, 5-16

D

Denmark Strait overflow,
  III.15, 1, 3, 4, 11-17

D cont.

Dense bottom current
  continuity eqt., 41, 97
  depth-variation, 41
  entrainment function, 96
  miscible, 81-98
  momentum eqt., 42-43
  non-miscible, 17-19
  Richardson number for, 93
Density
  distribution, 65
  of seawater, ρ, 157
  reduced, Δ, 159
Diffuser, 147-154, III.14, 1-10
Diffusion, 61
  transport by, 118
Dilution, 154, III.14, 4
Dissipation, rate of energy,
  53-118
Downwelling, 113-115

E

Earth rotation, III.15, 5-16
Ekman depth, 114-115
Energy
  consideration, 3-5
  dissipation, 53, 118
  gain in kinetic (KIN), 53
  potential (POT), 12-16
  production of turbulent
    kinetic (PROD), see
    Production ...
Energy equation
  free penetrative convec-
    tion, 106-112
  mean motion, 45
  salt water wedge, 20
  turbulence, 51-54
Entrainment, V_E, 36-54, 61-70
  buoyancy flux by, 61
  function for
    dense bottom currents, 96
    free penetrative con-
      vection, 112, III.16,1-8
    horizontal buoyancy flow,
      144
    jets and plumes, 150
    winddriven stratified
      flow, 124
  two-ways, 67
Environmental problems, 2-5